MW01223753

Tortillas and Tomatoes

McGILL-QUEEN'S STUDIES IN ETHNIC HISTORY
DONALD HARMAN AKENSON, EDITOR

1 Irish Migrants in the Canadas
A New Approach
Bruce S. Elliott

2 Critical Years in Immigration
Canada and Australia Compared
Freda Hawkins
(Second edition, 1991)

3 Italians in Toronto
Development of a National Identity, 1875–1935
John E. Zucchi

4 Linguistics and Poetics of Latvian Folk Songs
Essays in Honour of the Sesquicentennial of the Birth of Kr. Barons
Vaira Vikis-Freibergs

5 Johan Schrøder's Travels in Canada, 1863
Orm Øverland

6 Class, Ethnicity, and Social Inequality
Christopher McAll

7 The Victorian Interpretation of Racial Conflict
The Maori, the British, and the New Zealand Wars
James Belich

8 White Canada Forever
Popular Attitudes and Public Policy towards Orientals in British Columbia
W. Peter Ward
(Second edition, 1990)

9 The People of Glengarry
Highlanders in Transition, 1745–1820
Marianne McLean

10 Vancouver's Chinatown
Racial Discourse in Canada, 1875–1980
Kay J. Anderson

11 Best Left as Indians
Native–White Relations in the Yukon Territory, 1840–1973
Ken Coates

12 Such Hardworking People
Italian Immigrants in Postwar Toronto
Franca Iacovetta

13 The Little Slaves of the Harp
Italian Child Street Musicians in Nineteenth-Century Paris, London, and New York
John E. Zucchi

14 The Light of Nature and the Law of God
Antislavery in Ontario, 1833–1877
Allen P. Stouffer

15 Drum Songs
Glimpses of Dene History
Kerry Abel

16 Louis Rosenberg
Canada's Jews
Edited by Morton Weinfeld

17 A New Lease on Life
Landlords, Tenants, and Immigrants in Ireland and Canada
Catherine Anne Wilson

18 In Search of Paradise
The Odyssey of an Italian Family
Susan Gabori

19 Ethnicity in the Mainstream
Three Studies of English Canadian Culture in Ontario
Pauline Greenhill

20 Patriots and Proletarians
The Politicization of Hungarian Immigrants in Canada, 1923–1939
Carmela Patrias

21 The Four Quarters of the Night
The Life-Journey of an Emigrant Sikh
Tara Singh Bains and Hugh Johnson

22 Resistance and Pluralism
A Cultural History of Guyana, 1838–1900
Brian L. Moore

23 Search Out the Land
The Jews and the Growth of Equality in British Colonial America, 1740–1867
Sheldon J. Godfrey and Judith C. Godfrey

24 The Development of Elites in Acadian New Brunswick, 1861–1881
Sheila M. Andrew

25 Journey to Vaja
Reconstructing the World of a Hungarian-Jewish Family
Elaine Kalman Naves

McGILL-QUEEN'S STUDIES IN ETHNIC HISTORY
SERIES TWO: JOHN ZUCCHI, EDITOR

Inside Ethnic Families
Three Generations of Portuguese-Canadians
Edite Noivo

A House of Words
Jewish Writing, Identity, and Memory
Norman Ravvin

Oatmeal and the Catechism
Scottish Gaelic Settlers in Quebec
Margaret Bennett

With Scarcely a Ripple
Anglo-Canadian Migration into the United States and Western Canada, 1880–1920
Randy William Widdis

Creating Societies
Immigrant Lives in Canada
Dirk Hoerder

Social Discredit
Anti-Semitism, Social Credit, and the Jewish Response
Janine Stingel

Coalescence of Styles
The Ethnic Heritage of St-John River Valley Regional Furniture, 1763–1851
Jane L. Cook

Brigh an Òrain/A Story in Every Song
The Songs and Tales of Lauchie MacLellan
Translated and Edited by John Shaw

Demography, State and Society
Irish Migration to Britain, 1921–1971
Enda Delaney

The West Indians of Costa Rica
Race, Class, and the Integration of an Ethnic Minority
Ronald N. Harpelle

Canada and the Ukrainian Question, 1939–1945
Bohdan S. Kordan

Tortillas and Tomatoes
Transmigrant Mexican Harvesters in Canada
Tanya Basok

Tortillas and Tomatoes

Transmigrant Mexican Harvesters in Canada

TANYA BASOK

McGill-Queen's University Press
Montreal & Kingston · London · Ithaca

ISBN 0-7735-2338-3

Legal deposit second quarter 2002
Bibliothèque nationale du Québec

Printed in Canada on acid-free paper that is 100% ancient forest free (100% post-consumer recycled), processed chlorine free, and printed with vegetable-based, low VOC inks.

This book has been published with the help of a grant from the Humanities and Social Sciences Federation of Canada, using funds provided by the Social Sciences and Humanities Research Council of Canada. Funding has also been received from the Office of Research Services, the Department of Sociology and Anthropology, and the Office of the Executive Dean, Faculty of Arts and Sciences, of the University of Windsor.

McGill-Queen's University Press acknowledges the financial support of the Government of Canada through the Book Publishing Industry Development Program (BPIDP) for its publishing activities. It also acknowledges the support of the Canada Council for the Arts for its publishing program.

National Library of Canada Cataloguing in Publication Data

Basok, Tanya, 1958–
Tortillas and tomatoes: transmigrant Mexican harvesters in Canada
Includes bibliographical references and index.
ISBN 0-7735-2338-3
1. Alien labor, Mexican – Ontario. 2. Agricultural laborers, Foreign – Ontario. I. Title.
HD1530.05B38 2002 331.6'2720713 C2001-903085-1

Typeset in Sabon 10.5/13
by Caractéra inc., Quebec City

To Nathaniel

Contents

Tables and Figures

TABLES

FIGURES

Preface

In the early 1980s Smit, Johnston, and Morse (1984, 1–2) noted that, "Despite the importance of a secure labour supply to the maintenance of a healthy agricultural sector, the farm labour market, especially that for seasonal workers, has been difficult to investigate. More is known about most other aspects of the agricultural production system than about hired labour in general and about seasonal labour in particular." These comments are still relevant today. However, even less is known about foreign seasonal workers. This books aims to fill this gap.

Research for this book germinated in praxis. It began in 1997 when I was conducting field research on the economic impact of labour migration among Mexicans employed in the Leamington area through the government-run Commonwealth Caribbean and Mexican Seasonal Workers Program. That summer was unusual in Leamington. Two Mexicans were hit by cars in two separate accidents. One was killed, and the other spent several months in a coma in a Windsor hospital. As a result, Mexicans in the area began to think about what their lives were worth, that is, whether their families would have to bear the expense of transporting their bodies back to Mexico in the event of death and how much life and disability insurance coverage they had under the employment program. Some thought they were not insured at all; others thought their policies were worth only four thousand dollars; still others knew they were insured but did not know the exact value (which, in fact, was forty thousand dollars). Apparently some workers had even asked the consulate representative who came to visit them about it, and the representative was reported to have retorted, "Why do you need to know? Do you intend to die?" Some Mexicans asked me to find out whether they had life insurance coverage at all.

It was not until one Sunday afternoon that I realized how little Mexican seasonal workers were aware of their rights concerning working or living arrangements. Having finished interviewing Mexican farm workers at a soccer field, I went to say hello to some Mexicans I had interviewed the previous summer. One of them, after uttering a polite hello, took a belligerent stand: "Now that you have been doing research on us for a year, can you tell us what rights we have." Before I knew it, I was surrounded by some fifty Mexican men who, interrupting each other, bombarded me with questions:

- Are we entitled to overtime pay?
- Are we entitled to sickness pay?
- Do we have life insurance?
- Do we have insurance for work-related accidents?
- Why do workers in Quebec get 4 percent vacation pay and paid public holidays when workers in Ontario do not get these payments? Does each province have a different policy? Why do some of us get vacation pay here when others do not?

The uncertainty surrounding their working and living conditions creates much anxiety among the seasonal workers. I realized that while the agreement for the employment of Mexican seasonal workers had been signed by the Canadian and Mexican governments, the Mexican consulate was responsible for ensuring that the terms of the agreement were met and those responsible for the program made no effort to inform the workers themselves of their rights. The resulting ignorance with respect to their rights renders Mexican workers vulnerable.

Having read the employer information package provided by the Foreign Agricultural Resource Management Service (FARMS) to every grower employing offshore labour, I knew answers to some of the questions thrown at me, but I needed to do some homework to obtain information on the questions I could not answer. We agreed to hold a meeting at which I would attempt to provide information to the Mexican workers about their entitlements. We met twice, both times on Sundays, right after a service held in Spanish in the basement of the Catholic church where the service was held. The church priest gave us his approval, if not his blessing.

On the day when I was approached by the Mexican workers and at the two meetings held in the church basement after the Sunday service, I heard numerous accounts of illnesses, accidents, allergic and neurological reactions to pesticides, and unsafe pesticide spraying practices. I learned about one man who had lost his vision in one eye because he had been sprayed with a pesticide. I heard accounts of people working while pesticides were sprayed just a few metres away. I heard of Mexican workers having to spray pesticides when no protective clothing had been provided to them by their employers. And I learned that there was no running water at some work places, making it impossible for the workers to wash the pesticide off their hands before eating lunch.

Until then I had avoided asking questions about the working and living conditions of Mexican seasonal workers in Canada, not because I was not interested in them but merely in order to avoid getting my research participants into trouble with their employers. When I started conducting research among the Mexicans in Leamington in 1996, I had to rely on growers to introduce me to their Mexican workers, since I could not obtain lists of farms employing Mexican seasonal workers from any official sources. Claiming that this information was confidential, the Leamington employment office refused to help me. I had to use a contact with a local grower who gave me a few names and telephone numbers of farmers who hired offshore workers. And it was through these farmers that I met some Mexican workers. However, being introduced to them by the employer limited the range of questions I could ask during my semistructured interview. In order not to make my respondents suspicious and fearful that the information they offered me would jeopardize their jobs, I focused on the questions that were safe, that is, questions that were directly relevant to my interest in the economic impact of the Canada-bound migration and that were of no consequence to the employer-employee relationship. When I returned to the field in the summer of 1997, I followed the same approach, and my interview schedule omitted any dangerous questions. Yet, as the events described above clearly demonstrated, the workers themselves felt that it was important for us to talk about their working and living conditions. And therefore these issues were raised not only in the church basements but during many interviews that I and my research assistant, Nicole Noel, conducted that year in Leamington and later in the Mexican village of San Cristóbal.

In 1997 I used a different approach to locate Mexican workers that allowed me to cover topics related to employment and housing conditions, without making my interviewees nervous about this information leaking to their bosses. In the summer of 1997 I went to visit some workers among whom I had conducted interviews the previous summer. Nicole and I spent the first few weeks interacting with these workers informally – taking them shopping in Windsor, dropping in for a bite to eat, going out with them for lemonade, and giving them a few English lessons. Then I asked them to introduce me to Mexicans working on other farms. Sometimes I ran into farm owners when I came to conduct interviews; at other times, especially if the workers did not live on the farm premises, the farm owners were unaware of our interviews. In both cases, we were introduced to the workers by their fellow Mexicans, and not by the *patrones* (employers).

Soon after we started our fieldwork, we realized that the best day for interviews was Sunday, since most Mexicans worked only until noon, and that the best place for interviews was the soccer field. Weather permitting, Mexican workers would gather around the field while some played and others watched. Not having to worry about cooking dinner, eating, or washing clothes (things that many Mexicans have to do after their long days at work during the week), they were more willing to talk to us. We seized this opportunity and conducted many interviews while soccer games went on. We also used our soccer field acquaintances to lead us to their farms, where we could interview their coworkers. This form of locating respondents, which by-passed an introduction from the employer altogether, led us to some topics that Mexicans themselves found important. Among other things, we started talking about their working conditions in Canada. However, because we still waited for our research participants to volunteer this information, the research data gathered at that stage was largely anecdotal.

It was only when Nicole and I went to San Cristóbal in the state of Guanajuato, Mexico, that we addressed the issues of working and living conditions in a systematic way, by including the relevant questions in our interview schedule. San Cristóbal has more participants in the Canadian Agricultural Seasonal Worker program than any other Mexican village. By going to San Cristóbal I hoped to interview participants in the program who were employed outside the Leamington area, former participants who had stopped coming to Canada, and Mexicans employed in the Leamington region who were not among the 154 workers we had located and interviewed in 1996 and during the summer and fall of 1997. I continued gathering information on the use of Canadian earned remittances but added to my interview schedule several questions that dealt specifically with the employment and housing conditions Mexican workers experienced in Canada. Thus my research participants redefined my research project, adding another dimension to it that they themselves found worthy of attention. But this book is not just about conditions under which Mexican seasonal workers are expected to work in Canada. It is also about why they put up with job requirements they find demanding and even abusive. My aborted activism provided answers to this question.

Much to my frustration, my attempts to solve the problems of Mexican workers bore no fruit. The first meeting in the church basement was attended by some two hundred workers. But only some fifty of them showed up for the second meeting, when two representatives of the Windsor Occupational Health Information Services came to talk to them. At the first meeting I asked the workers, who could not stop giving me examples of abuses they had suffered, to make a list of farms and indicate whether their *patrones* offered them vacation pay and public holiday pay and whether they used unsafe work practices. Not a single person volunteered to make such a list. At the second meeting, those who attended promised to look up the labels on the pesticide cans used on the farms and give me the names of the pesticides when they saw me next, but I did not receive a single name. My informants did not wish to be known by their *patrones* for their rebellious behaviour, and they did not trust each other enough to hope that their activism would be kept secret from them. Even when they talked openly about the problems they faced at work or the inadequacy of their living quarters, as soon as they were asked to provide the names of their *patrones*, their response was that it was others who had these problems and that their *patrón es buena gente* (is a good guy).

My first reaction was disappointment and even anger at the Mexican workers who had made me do a lot of work on their behalf and then lost interest in our project. I felt in some way betrayed.

But then I realized that the workers were afraid to jeopardize their chances of returning to Canada by demanding that their rights be respected. Of course they wished their working conditions would improve. Of course they wanted to receive their 4 percent vacation pay and paid public holidays if they were entitled. But they were not prepared to engage in any political activism, for fear of being excluded from the program. This experience made me think about mechanisms of social control that assured the compliance of Mexican workers with the working and living conditions they experienced, no matter how abusive they may have been.

Another question that was on my mind was whether Mexican workers were displacing domestic labour. To collect information that the Mexican workers requested, I met with staff members from the Windsor Occupational Health Information Service. A local labour activist was invited to our meetings. He was very sympathetic with the plight of Mexican workers and, in fact, helped one injured worker to get a cornea transplant. But his major interest was to "get" the big greenhouse growers who, he felt, made tremendous profits on the backs of their workers. His position was that these growers could easily offer better working conditions and increase the wages they paid their workers, which should have made the jobs attractive to domestic workers. I discussed these concerns with Nicole. As a graduate student she had often had to accept jobs paying a minimum wage, and she knew many other students who were desperately trying to get a job and who would, she felt, accept the pay and the working conditions that farmers were willing to offer. She too felt that the interests of local workers were undermined by the importation of Mexicans.

Having learned about the nature of farm work from my Mexican informants, I doubted that even if the wages were higher, there would be enough Canadian residents who would agree to work under the conditions that Mexicans accepted – with long hours, no days of rest, extreme heat, and no social life outside work. I felt that I needed to hear from local growers about the experiences they had had with local workers and thus originated the other part of the project that led to this book – interviews with forty-five Leamington greenhouse growers. I focused on greenhouse growers for two reasons. First, the greenhouse industry has been experiencing tremendous growth in the last decade. I felt that if there was room for any improvements of working conditions for farm workers and increases in the levels of remuneration needed to attract local labour, it would be in the greenhouse vegetable sector. The second reason was practical. The city of Leamington economic development unit had compiled a directory of greenhouse growers with their names and phone and fax numbers. My other research assistant, Carolyn Lewandowski, used this directory to approach greenhouse growers and ask them for interviews. At the same time, the Ontario Fruit and Vegetable Growers Association refused to provide us with a list of other Essex growers. So we had to make do with what we had. Yet many issues that greenhouse growers raised in their interviews can be easily extended to other growers. Interviews with Leamington greenhouse growers pointed out that offshore labour was vital for the survival of the Ontario farming industry, and particularly its greenhouse sector, and that without these workers many Ontario farms would close down. Therefore, my book also deals with conditions in Ontario farming that make unfree migrant labour indispensable.

The book reverses the order in which the various issues discussed above were addressed in my research project. I begin the book with the discussion of Ontario farming, the problems growers have had with local labour, and the reasons they require workers who would be fully committed to the job for as long as the harvest period lasts. In the second half of the book, by explaining their willingness to work long hours virtually every day of the week, even when sick or injured, I argue that Mexican seasonal workers are the ideal farm labour force. The major argument of the book is that Mexican seasonal workers have become a "structural necessity" in many rural Ontario communities such as Leamington and that without offshore labour Ontario horticulture would have experienced a significant decline.

It would have been impossible for me to compile the data for this book without the generous financial assistance of the Association of Universities and Colleges of Canada, the International Development Research Centre, the University of Windsor Research Board, and the Social Science and Humanities Research Council. I am also grateful to many people, including Mexican workers, greenhouse growers, officials at the Mexican consulate in Toronto, and employers at the Mexican Ministry of Labour and Social Planning, for spending their time with me and my research assistants so that we would gain a good understanding of this guest worker program.

To Nicole Noel, my research assistant, who has participated in all stages of this research project, from 1997 until 2000, I owe deep gratitude, not only for her outstanding research skills but also because she was a wonderful travel companion. Other research assistants, Carolyn Lewandowski, Jessica Cattaneo, and Adriana Prevat, were also instrumental to this project. I would also like to thank my son, Nathaniel, for putting up with all the inconveniences I forced him to endure while conducting the field work and for remaining cheerful most of the time. And finally, I am grateful for the comments made by two anonymous readers on the earlier draft of this book.

Tortillas and Tomatoes

1 Migrant Labour: A Structural Necessity?

Leamington-area residents know that it is better to avoid the No Frills grocery store on late Friday afternoon, when about one thousand Mexican seasonal farm workers employed in the area go shopping there. For Leamington-area residents the Mexican invasion of the local supermarket has become a part of the social landscape, as has the image of Mexican men riding their bicycles along rural roads, particularly noticeable on Sunday afternoon when most Mexicans get time off work. Over the last twenty-five years, the presence of these men, who look and dress differently from much of the local population and who spend between five and eight months a year in this area, has become more and more noticeable as the number of Mexican seasonal workers has progressively increased. The rapidly expanding greenhouse vegetable industry, which is highly dependent on the recruitment of foreign workers, has been largely responsible for the increase in the Mexican male population. Mexican seasonal farm workers have become, to use Cornelius's (1998) term, "structurally embedded" in the economy of the region. Two central questions are therefore addressed in this book. First, what makes it necessary for Ontario growers to hire offshore workers? And second, why are Mexican men who do seasonal work in Ontario so ideally suited for the job? To put it in more general terms, the first question concerns the demand for foreign labour and the second addresses the workers' compliance with job requirements. The first question has been addressed in the literature on migrant labour. However, as will be seen below, this literature fails to distinguish foreign labour that is a "structural necessity" from foreign labour that is merely preferable and convenient.

I will argue in this book that migrant labour is structurally necessary only in those economic sectors whose viability hinges on the employment of workers who are *unfree*. Robert Miles defines unfree labour as workers whose ability to circulate in the labour market is restrained through political and legal compulsion (1987, 32–3). In this book I expand the definition of unfree labour by including the workers' inability to refuse employers' demands. In other words, unfree workers are not only unable to change employment, but they are also unfree to refuse the employers' requests for their labour whenever the need arises. Of course, slavery of this kind does not exist in today's world. However, temporary recruitment of contract workers among those whose economic needs force them to accept the terms of unfreedom comes close to it. While seasonal migrant workers are legally free to quit or to refuse labour to their employers, economic pressures that force them to accept seasonal contracts in a foreign country, combined with the mechanisms of control inherent in a government-regulated recruitment program, make them more willing to accept their conditions of unfreedom.*

DEMAND FOR FOREIGN LABOUR

In the political-economy paradigm, demand for foreign labour has been explained from three perspectives. The first one, pioneered by Castles and Kosack (1985), relates employment of foreign labour to absolute shortages in the "relative surplus population." The second, or the segmented labour-market approach, links the employment of foreign workers to jobs in vulnerable economic sectors that pay low salaries and offer dismal working conditions, rendering them unattractive to domestic workers (Piore 1979;

* Even though the application of the concept of unfree labour to Mexican and Caribbean workers in Canada is not unproblematic (for a detailed discussion of this issue see Basok 1999 and Knowles 1997), I chose to use it in this book because it captures well the constraints imposed on Mexican seasonal workers that render them ideal for Canadian growers.

Cohen 1987). And finally, the global restructuring perspective associates jobs held by (im)migrants with poorly paid and insecure jobs found in both vulnerable and dynamic economic sectors (Sassen-Koob 1985; Fernández Kelly 1985; Nash and Fernández Kelly 1983). These three approaches will be reviewed below.

Foreign Workers as Relative Surplus Population

The first approach to understanding the importation of foreign labour was formulated to explain the labour recruitment program in Western Europe during the couple of decades following World War II. For several reasons, during this time Western Europe experienced an economic boom marked by high rates of capital accumulation (Castles, Booth, and Wallace 1984, 20–2). In a situation of full employment, the industrial reserve army that is "the lever of capitalist accumulation" and "a condition for the existence of the capitalist mode of production" (Marx 1976, 786) has disappeared. Marx distinguishes three groups within the "reserve army," or the "relative surplus population." First, there is the latent reserve population, comprised of those who have been displaced by the penetration of capitalist relations of production into agriculture. Second, there is a floating surplus population consisting of those who have been expelled from the process of production by the operation of the law of capital accumulation. And third, there is a stagnant surplus population, such as disabled, sick, and homeless people (794–8). None of these groups of people could provide sufficient labour to keep up with the rates of accumulation in postwar Europe (Miles 1987, 145). The reserve army of the unemployed, or the floating relative surplus population, was reduced by the economic expansion. Agricultural reforms caused a significant decline in the agricultural population, making it impossible to recruit among the latent surplus population. Neither could women and others in the stagnant surplus population satisfy the demand in the labour force (Castles, Booth, and Wallace 1984, 24–6). Cohen lists three alternatives available to Western European countries that failed to recruit among the "relative surplus population": increasing the rate of exploitation of the existing labour force by raising productivity and extending working hours, investing in labour-saving technology, and relocating to regions where labour costs were lower. According to Cohen these three alternatives also had limitations in Western Europe at that time (116–22).

Foreign workers were recruited to fill positions in the lower echelons of the occupational ladder. In a situation of full employment native workers moved into white-collar or skilled occupations that enjoyed better pay and were less difficult and dangerous. Having had an opportunity to receive vocational training, native workers were reluctant to remain in unskilled manual jobs, thus creating demand for labour in the occupations that they had deserted (Castles and Kosack 1985, 26–7).

The recruitment of foreign labour did not merely fill the gaps in the labour market; it also offered several other advantages to capital and the state. Among them were the following. First, in a receiving social formation, capital and the state could avoid the reproduction costs of foreign workers who left their families behind and who were expected to return home themselves once the need for their labour subsided. Second, foreign workers were not only cheaper, but they were also flexible, less organized, prepared to work unsociable hours, and easier to hire and fire, and they had lower economic and social expectations. And third, racial and national divisions in the labour force fragmented the Western European working class (Cohen 1987, 123–40; Castells 1979, 361–72; Castles and Kosack 1985), making it easy for capital to control it.

Foreign Workers and the Segmented Labour Market

Segmented-labour theorists have linked employment of (im)migrant labour to segmentation in the labour market that is related to the vulnerability of some economic sectors that offer poor working conditions and low pay. According to segmented labour-market theorists (Edwards, Reich, and Gordon, 1973; Gordon, Edwards, and Reich, 1982), the homogenization of labour that accompanied the development of competitive capitalism in the nineteenth century came to an end by that turn of the century when capitalist development underwent a significant transformation to monopolistic capitalism. In the system of monopoly capitalism, giant oligopolistic corporations dominate the economy. They coexist with a surviving peripheral capitalist sector comprised of industries that face cyclical, seasonal, or otherwise unstable demand (Edwards, Reich, and Gordon, 1973; Gordon, Edwards, and Reich, 1982). These unstable economic sectors are often under intense competitive pressure from abroad, but insecurity can also be related to seasonal fluctuations and fluctuations in the weather, changes in fashions and trends, or market cycles (Piore 1979, 36; Castells 1979, 366; Cohen 1987, 128). Labour-intensive technologies and lack of market power restrict the ability of these peripheral sectors to pay higher wages (Harrison and Sum 1979, 690). This structural dualism corresponds to two labour markets – the primary and secondary markets.

Primary-sector jobs are characterized by stability, while jobs in the secondary sector are unstable (Edwards 1973, xi–v). While primary sector jobs are more attractive and better paid, secondary sector jobs are associated with relatively low wages, low skill requirements, poor working conditions, limited possibilities for upward mobility, and the generally inferior or demeaning social status that is attached to them (Piore 1979, 17; Berger and Piore 1980, 17–18). According to Edwards the primary market includes workers "who follow a logical progression from job to job, in which pay, responsibility, authority, and status increase with labour force experience," and the secondary market is comprised of workers "who follow a much more random series of jobs and are generally denied opportunities for acquiring skills and for advancement" (Edwards 1973, 16). Furthermore, while labour in the primary sector is subject to bureaucratic control, in the secondary sector old practices of "simple-hierarchy" control, characterized by harsh, personalized, and often arbitrary exercise of power, are used (Edwards 1973, 20; Piore 1979, 17; Berger and Piore 1980, 18). Harrison and Sum have suggested that low wages and the virtual absence of fringe benefits, combined with undesirable working conditions, produce high turnover in the labour force in the secondary sector, and, in turn, the lack of stability impedes unionization. While in the primary sector there are well-defined career ladders, making promotion more predictable, secondary labour markets are dominated by firms whose "internal labour markets" lack this structure (1979, 690).*

Relatively good working conditions in the primary sectors have been attributed to the achievements of organized labour. Having

* Piore (1973) elaborated the original dualistic model by suggesting that the primary sector is subdivided into an upper and lower tier. Workers in the upper tier tend to have more job control than workers in the lower tier, where tasks tend to be more routinized.

formed the Congress of Industrial Organizations (CIO) in the 1930s, North American workers fought hard for recognition of collective bargaining, grievance procedures, and seniority rules for allocation of work. But industrial unionism came to be concentrated in large firms that could afford to pay workers well and offer them security. By contrast, smaller peripheral firms could offer only nonunionized, low-paid, part-time, casual, and unstable jobs to its workers (Gordon, Edwards, and Reich, 1982, 190–200), because of the instability that they themselves faced due to intense competitive pressure from abroad or due to cyclical or seasonal demands. If these industries were to pay higher wages, they could not survive (Piore 1979, 28; Cohen 1987, 128). Since many native workers find secondary sector jobs unattractive, economic sectors that offer them become dependent on foreign labour (Boswell and Brueggemann 2000).

Foreign Labour and Global Restructuring

Global restructuring theorists have pointed out that as a result of global restructuring many new and old jobs in the primary sector are poorly paid and insecure and therefore unattractive to domestic workers with middle-class aspirations. These jobs are as likely to attract (im)migrant labour as jobs in the traditional secondary sector. It has been argued that global restructuring redefined the boundaries of the segmented labour market, moving many jobs characterized as part of the secondary sector into the primary sector. In the new global structure "low-wage dead-end jobs can be part of highly dynamic, technologically advanced growth sectors" (Sassen-Koob 1985, 231).

Several decades of postwar prosperity came to a halt in the 1970s when private capital accumulation rates began to decline. This crisis provoked global restructuring, which aimed to reestablish the foundations for sustained profitability and accumulation (Warrian 1987, 20; Carroll and Warburton 1991, 323). One of the responses to this crisis was disinvestment, or a diversion away from productive investment in basic industries such as the steel, automobile, electrical, and textile industries, among others, and into unproductive speculation, mergers and acquisitions, and foreign investment (Bluestone and Harrison 1982, 6) . Businesses have used such practices as vertical disintegration, downsizing, outsourcing, and the formation of networks of companies that operate globally and across sectoral boundaries to minimize their production costs (Harrison 1994, 196). The relocation of production has been facilitated by advances in industrial, transportation, and communications technologies, making it possible for transnational corporations to establish parallel production and numerous instances of sourcing on a global scale (Warrian 1987, 22). Some researchers have referred to this situation as the new international division of labour, to distinguish it from the "old" internationalization. Whereas in the past, local labour was used by foreign corporations in developing countries to exploit local resources, since the early 1960s it has been the manufacturing stages of production that have been relocated to these countries (Fernández Kelly 1985, 210).*

Under these new conditions the association between backward jobs, backward sectors, and the process of decline is undermined (Sassen-Koob 1985). Poorly paid and insecure employment is now characteristic of three areas: the service sector, the downgraded old industries, and newly emerging industries. As a result, insecure, poorly paid jobs, often filled by (im)migrant workers, whether legally employed or undocumented, are found in a wide range of occupational areas. Among them are domestic services, brickworks, foundries, construction, tanneries, coal mines, agriculture, hotels and restaurants, hospitals, railways and other public transport, assembly work in manufacturing, and the garment industry (Sassen 1996; Harris 1995, 35–48, 175–7; Stalker 2000, 132–4).

The creation of low-paid employment in the service sector has been linked to the emergence of "global cities," which acquire a new role in the international division of labour as producers of management and control functions (Sassen-Koob 1985). Several activities, such as finance, insurance, real estate, and business services are associated with this new role that these cities assume. These new areas of production generate, directly or indirectly, a significant supply of low-wage jobs filled by (im)migrant labour. And rapid growth in the number of financial experts, technical

* Fernández Kelly traces the beginning of this process to 1962, when Fairchild, the electronics manufacturer, opened the first offshore semiconductor assembly plant in Hong Kong. In 1966 Fairchild opened a manufacturing plant in Korea, and in 1964 General Instruments moved its microelectronics assembly operations to Taiwan (206).

consultants, and international lawyers stimulates expansion in other areas of the economy, which range from office work to paper and software production, and which in turn lead to an expanded need for cleaners, stock clerks, and auxiliary workers. Thus, the growth of highly specialized services associated with high incomes also stimulates demand for cheap labour to service the activities and life-styles of high-income strata.

The downgrading process includes three trends: the social reorganization of the work process, such as the expansion of sweatshops and industrial homework, particularly in the garments and electronics industries; the technological transformation of the work process; and the rapid expansion of high-technology industries that offer mainly low-paid employment (Sassen-Koob 1985, 238). The reorganization of the labour process that is designed to intensify work and raise productivity includes measures such as splitting work processes into simple operations to reduce the need for skilled workers and the increased use of assembly lines, piece-work, shift work, and bonus payments (Castles 1984, 27; Lipsig-Mammé 1987, 43–4). As Fernández Kelly (1985, 217–18) points out, even most technologically advanced sectors have generated their own types of sweatshops. Various telecommuting offices now rent equipment to predominantly female workers to be used at home to process various kinds of information. At home they are paid by piecework or job lot; they are unprotected by laws related to minimum wages, unemployment insurance, workers' compensation, overtime, paid breaks, sick leave, or paid holidays, and they receive no fringe benefits like health insurance (Katz and Kemnitzer 1983, 339).

In the 1980s, as a result of global restructuring, many concessions won by organized labour since the 1920s were lost. The new post-Fordist labour regime has been referred to as "hegemonic despotism." Labour relations are still mediated by the state, as in the Fordist days, but, as Carroll and Warburton (1991, 323) put it, "where labour used to be *granted* concessions based on the expansion of productivity and profits, it now *makes* concessions in the international competition for investors" (italics in the original). The state has supported capital by intervening in collective bargaining, not only in the public sector but also in the private sector, by introducing back-to-work legislation for public workers and imposing restraints on wage negotiations (Panitch and Swartz 1998).

Related to these developments is the drop in levels of unionization (Nash 1983, 27) As Bluestone and Harrison (1982, 179) point out, even in those industries where unions felt secure, management began to openly threaten plant shutdowns unless the unions agreed to reopen contracts and accept wage freezes and reduced benefits and protections. Furthermore, the rate of decertification of prior union victories went up significantly in the 1970s (Bluestone and Harrison 1982, 179). Deunionization is sharpest in areas with rapid growth in high-tech industries (Sassen-Koob 1985, 238). Fernándes Kelly (1985, 215) points out that electronics industries are notorious for their resistance to unionization drives.

Most corporations are no longer interested in accommodating the interests of organized labour. They push for wage concessions and the end of collective bargaining. In this new labour market, new jobs offer predominantly part-time, low-wage, low-benefit, and unstable employment (Warrian 1987, 26; Fernández Kelly 1985, 214–15; Harrison 1994, 197). In the United States in the late 1980s as much as a quarter to a third of the civilian labour force was comprised of "contingent" workers, or workers employed in part-time, temporary, and contractually limited jobs (Harrison 1994, 201).

Under the new labour relations regime working conditions have deteriorated significantly for many workers who had enjoyed stability and prosperity. "Lean and mean" production generates a new dualism in the labour market that is characterized by the shrinkage in the number of safe, stable, and secure positions, even among white-collar workers. The drive for flexibility and cost savings has been facilitated by the growth of an educated, more diverse urban labour force and the standardization of office and other white-collar work in North America (Harrison 1994, 199). Warrian observes that falling real wages, layoffs, and a loss of rights for primary labour market workers "is resulting in a reshuffling of the boundaries of privilege within the working class, such that all workers may become 'secondary'" (1987, 26). When jobs move oversees, the unions' ability to allocate preferred jobs to white males is undermined (Nash 1983, 31; Jaret 1991). However, old divisions among workers do not disappear. If anything, the process of restructuring may deepen these divisions (Carroll and Warburton 1991, 325). Katz and Kemnitzer (1983, 343) observe that "the powerless sectors – women, the very poor, nonwhites generally, and recent and/or undocumented (im)migrants in particular – seem once again played off against the more entrenched, traditional, industrial working class."

The downgraded jobs are often filled by (im)migrant workers. While domestic workers tend to resist restructuring measures, (im)migrant workers are quite willing to comply with the changes (Castles 1984, 27). As Sassen (1988, 26) observes, "The contradiction between the existence of labour shortages and the existence of large unemployment worldwide is due to various factors which generate differentiation within the labor supply and in the demand for labor. Among these factors are the price of labor, the expectations of workers, the need for certain types of economic systems to secure cheap and *docile* workers, and the technological transformation of the work process in the last ten years that has not only upgraded some jobs but also downgraded many more jobs, making them unattractive to workers with middle-class aspirations."

In the United States Gonzalez, Baker, et al. (1998) have illustrated that Mexican workers became attractive to employers when old unionized plants closed down and that new employment arose mainly in nonunionized shops. They suggest that because foreign-born workers in U.S. firms are less likely to advance to positions requiring education, English proficiency, and legal status than are U.S.-born workers, "a new tier of low-paid workers consolidates in the lower echelons" (91). With respect to immigrant women, Morokvasic (1984, 886) observes that "Women from the peripheral zones ... represent a ready made labor supply which is, at once, the most vulnerable, the most flexible and, at least in the beginning, the least demanding work force. They have been incorporated into sexually segregated labor markets at the lowest stratum in high technology industries or at the 'cheapest' sectors in those industries which are labor intensive and employ the cheapest labor to remain competitive."

Lipsig-Mammé (1987, 44) points out that immigrant women are hired not only in those sectors "where conditions and security are undermined because the industry, facing international competition, is desperate to cut costs, and does so on the backs of its workers" but also in sectors "where international competition is not an immediate factor, but where employers have borrowed tactics from the internationally vulnerable industries to cut their labour costs, reduce their labour force, augment 'flexibility' and weaken the union presence." Other researchers have demonstrated that new high-tech industries employ large numbers of female immigrant workers for assembly line jobs (Sassen-Koob 1985, 238; Green 1983). Harrison (1994, 117) depicts high-tech industry in the Silicon Valley as a "classic dual labor market," with technical and managerial jobs at the top and low-paid, expendable workers, predominantly of Latino and Asian origin, at the bottom, performing more mundane, standardized, and, at times, dangerous tasks.

EXPLAINING STRUCTURAL NECESSITY

None of the three approaches discussed above distinguishes between (im)migrants who are "structurally necessary" in a host economy and those who can be replaced by other categories of workers, such as unemployed, resident visible minorities, women, students, and disabled people. In fact, in the circumstances outlined by these three perspectives, foreign workers appear as a "preferred and helpful solution" for capital (Cohen 1987, 144) and not as a "structural necessity." In the Western European case discussed by the "relative surplus population" theorists, historical conditions changed by the early 1970s to render foreign workers expendable. First, the economic boom came to a halt by 1973. As rates of unemployment went up, it became possible to recruit among the domestic surplus population. Second, it became costly to socially reproduce foreign workers who had been joined by their families. And third, the price of foreign workers increased when foreign workers, supported by national trade unions, began to organize strikes and engage in other forms of political resistance (Cohen 1987, 140–3).

With respect to the segmented labour-market theory, radical economists representing this approach never claimed that (im)migrant workers were the only ones found in secondary-sector jobs. Other minorities, such as native-born blacks, women, disabled people, and youth, were said to occupy these jobs as well (Edwards, Reich, and Gordon 1973, 16; Waddoups and Assane 1993; Wilkinson 1991; Boston 1990). The employment of (im)migrant workers, rather than members of other minority groups, appears to be accidental in this model. The same argument applies to the global-restructuring perspective on the employment of foreign labour.

I would argue that because of their failure to distinguish between immigrant and migrant labour and free and unfree workers, these perspectives cannot explain the conditions that make the employment of foreign labour structurally necessary. Satzewich defines migrant labour as "those foreign born persons who seek to relocate themselves in sites in production but whose work and stay within a social formation are subject to temporal constraints imposed by the state" (1991, 38). Migrant workers are not granted the right of permanent settlement by the receiving state, and since they are not granted the rights of citizenship, they are deprived of the other liberal-democratic rights. By immigrant labour Satzewich means "those foreign-born persons who are awarded the right of permanent settlement" (39). Following Miles (1987), Satzewich distinguishes unfree labour from free labour by its inability to circulate in the labour market (1991, 42). Although unfree labour is anomalous in a capitalist mode of production, it is nevertheless necessary under certain historic circumstances "when labour markets prove unable to 'deliver' labour power at a price which permits the creation and appropriation of surplus value," making it necessary to recruit and retain labour by the use of political and legal compulsion and constraint (Miles 1987, 199). Because the liberal democratic state extends freedom-of-movement rights to its residents, legal immigrants cannot be employed as unfree labour. I would argue that migrants would be a structural necessity only if they were hired as unfree labour merely because no other categories of people, be they citizens or immigrants, could be employed under these conditions in a capitalist society. Both free migrants and free immigrants, although convenient, can be replaced by domestic labour if native workers are desperate enough to accept the "(im)migrant jobs" or if working conditions and pay are improved to meet the expectations of the domestic workers.

WORKERS' COMPLIANCE

As I mentioned earlier, I define "unfree" workers not only by their inability to change jobs but also by their willingness to provide labour whenever a need arises. The question that needs to be addressed is what makes certain categories of foreign workers willing to accept such conditions of captivity.

As many researchers have pointed out, economic need forces many people to seek employment abroad and to accept the exploitative working conditions they experience (Bonacich 1972; Piore

1979; Cohen 1987). However, free (im)migrant workers are free to refuse some employers' demands, even at the risk of losing a job, because they have an option of finding another job. This option is not available to unfree workers who are bound to a specific employer by a contract. The controlled nature of their recruitment, their fear of losing an opportunity to participate in the employment program, makes unfree workers acquiescent. Deportation and blacklisting from further participation in the temporary-worker program is one of the mechanisms of control used to penalize guest workers who breach the terms of their contract or who do not meet their employers' demands and expectations (Phizacklea 1983; Wood and McCoy 1985, 138–9; Satzewich 1991, 114; Howell 1982, 135; Tuddenham 1985). As Tuddenham points out, "The employer's control over the visa effectively places the worker in a state of economic peonage: the worker maintains his or her legal status at the sufferance of the employer, who determines whether the worker returns again the next year. To the extent the worker wishes to retain legal status, he is bound to do as the employer wishes" (39). Contract workers who are certified to work in specific jobs represent what Griffith (1986, 881) has referred to as a "captive labour force." They stay with the same employer for a pre-determined period (Howell 1982, 136). The vulnerability of seasonal workers makes them akin to illegal migrants who are considered by some authors to be the most vulnerable and exploited (Portes 1978; Castles 1984, 31–2). Yet numerous writers have pointed out advantages of contracted workers over illegal migrants. Wood and McCoy (1985, 138) contend that Caribbean cane-cutters recruited through the temporary foreign farmworker program known as the H-2 (currently H-2A) program to work in Florida, are equally, if not more, vulnerable to deportation than illegal workers. Questioning the commonly held assumption that undocumented workers are docile and easily exploited, Tuddenham (1985, 38) observes, "The rigors of migration and undocumented status tend to self-select highly motivated and sophisticated workers who know their own worth, and demand to be paid accordingly. Fear of [the] Immigration [and Naturalization] Service does not seem to alter this: an undocumented worker is as likely to be reported to INS by a low-paying employer as by a high-paying employer, so he may as well seek the higher-paying job. When an employer threatens to report a worker to INS, the worker is more likely to move on than agree to stay for lower wages."

By contrast, contract workers are not free to change employers. Calavita (1992, 58) observes with respect to *braceros*, seasonal farm workers imported to the United States from Mexico between 1942 and 1964, that "The bracero, operating outside of the free labor system, contracted for short periods of time, and delivered to the employer to do specific tasks as the need arose, provided an important element of predictability, stability, and – above all – control, in what was otherwise an unpredictable production process." She points out that working conditions in some crops (such as cotton) were so arduous that illegal immigrants often deserted their jobs. The *bracero* workers, on the other hand, were "*confined by law* to a given crop and employer" (56, italics in the original). Delgado cites a cane grower who said, "We used to own our slaves; now we rent them" (1983, 27).

Contract workers are not the only reliable workers. Other (im)migrants have been reported to be highly reliable, punctual, and available to work overtime, on weekends, and during night shifts, if needed (see Cornelius 1998, 126, for instance). Yet on occasion free (im)migrant workers choose to honour their family obligations over their jobs. They may also decide to return to their home countries or simply move to another job. By comparison with free workers, contract labour offers a higher degree of predictability, because it is bound to the employer by institutions that can easily expel disobedient workers from the employment program that migrants consider to be far superior to other options available to them.

CONTRACT, UNFREE LABOUR

OUTLINE OF THE BOOK

Two central arguments are made in this book. The first is that Ontario fruit and vegetable growers require not merely labour that is cheap but labour that is unfree – unfree to circulate in the labour market and unfree to refuse to work when required. Several researchers have argued that the recruitment of foreign agricultural workers is necessitated by farm labour shortages. These shortages arise because of the low pay and poor working conditions offered to agricultural workers. It has been argued that Canadian family farms, which are squeezed by the low food prices maintained by the Canadian state and increasing input prices, cannot afford to increase the pay and improve the working conditions of the farm labour they hire. At the same time, in its effort to protect family farms from ruin, the Canadian state prohibits organization of farm workers, rendering them unable to change their employment conditions (Satzewich 1991; Dawson and Freshwater 1975; Wall 1994; Parr 1985; Mitchell 1975; Stultz 1987; Shields 1992).

Viewed from this perspective, the demand for cheap agricultural labour is linked to the vulnerability of family farming. However, this perspective ignores the fact that some agricultural sectors are not vulnerable but have enjoyed stability and growth. The greenhouse vegetable sector is one of them. The high profits made by many growers in this sector should make it possible for them to increase workers' salaries in order to improve the rate of labour retention. However, while decreasing the turnover rate, higher salaries would leave other problems unresolved. Those workers who stay on the job are not willing to be chained to the job. Many have family obligations, church commitments, friendship ties, and personal needs that require them to take time off work. Yet during the harvest season, the crops cannot wait for the workers to return to work. When it is very hot, much of the harvest can be lost if workers do not show up; the growers need their workers to be available when the produce is ripe. And farms that have experienced growth and expansion do not wish to lose their crops due to shortages of available labour any more than small family farms. Mexican workers, who are recruited by the Mexican Ministry of Labour and allocated to the growers by the Canadian state, are an answer to these growers' prayers. Mexican workers are not necessarily cheap, but they are unfree. Their unfreedom is what the growers mostly like about these foreign workers.

The second argument is that the Mexican workers comply with their conditions of unfreedom – they stay until their contract expires; they do not take time off work, even when they are exhausted, sick, or injured, because their recruitment into the employment program is rigidly controlled. Their failure to meet the growers' expectations would result not just in a loss of their particular employment but in the loss of an opportunity to earn a decent income outside their country. In this sense they are different from free workers, even those who are found in the so-called secondary-sector jobs. The latter can move from one job to the next, even if the next one is as bad as the previous one. Contract workers do not have this option and are therefore much more likely to accept their conditions of unfreedom. Furthermore, having left their families behind, these workers are "chained" to the job, because they are unencumbered by any social obligations.

In part 1, I explore the reasons that impel Ontario farm and vegetable growers to hire foreign seasonal workers. Since the 1940s, land holdings in Canadian agriculture have been undergoing a significant degree of consolidation. As a result the consolidated farms relying on the labour of fewer farm operators and fewer household members require much more hired seasonal workers, despite a degree of mechanization. Shortages are felt particularly in the tobacco and fruit and vegetable industries and in the regions where these commodities are grown. Before the introduction of the so-called offshore program, the Canadian government had attempted to recruit Canadian labour through various programs that had failed to deliver reliable workers to the growers. Growers, desperate for farm help, appealed to the Canadian authorities to allow them to import Caribbean workers during the months of high demand. Under pressure from various growers' groups and the politicians who supported them, the Commonwealth Caribbean Seasonal Agricultural Program, known as the offshore program, was introduced in 1966, and in 1974 a similar program was created for Mexico. In Canada the program is administered by the Foreign Agricultural Resource Management Service (FARMS), which submits the request for Mexican workers to the Mexican Ministry of Labour and Social Planning. The latter selects eligible workers from the pool of applicants. Employment conditions for Mexican workers are regulated through formal agreements signed by representatives of the two governments, and employers are bound by the terms of these agreements. These issues are discussed in chapter 2.

Chapter 3, which is based on interviews with forty-five Leamington greenhouse growers, discusses the problems that these producers have experienced with local labour. First, it is hard to recruit local workers for farm jobs. Second, some of those who accept the jobs do not keep them. And third, those who do keep their jobs lack discipline and commitment to the farm. These characteristics make local workers unreliable in the eyes of the growers. As I illustrate in chapter 3, what most greenhouse growers want is not necessarily cheap labour but captive labour, or workers who are available whenever the crop is ready to be picked, including in the evening, on weekends, and on holidays. And they would be excluded from taking time off to fulfil any religious or family commitments. No Canadian workers (whether immigrants or native-born Canadians) are prepared to comply with these expectations.

Chapter 4 explores the reasons why Canadian tobacco and fruit and vegetable growers experience shortages of reliable labour. The chapter presents an explanation that links high turnover among farm workers to the low pay they receive and the dismal working conditions they experience on the farms. It also examines an argument made by various Canadian researchers that Canadian farmers cannot improve workers' pay and offer them any other benefits, because the farmers find themselves squeezed by rising agricultural input prices, high interest rates charged by the banks, and international competition, combined with a cheap food policy pursued by the Canadian state. This explanation portrays all Canadian farms as vulnerable. However, Canadian farming has experienced a significant degree of social differentiation. While some small family farms have been rendered extremely vulnerable, other farms have grown bigger and stronger. Chapter 5 focuses on the Leamington greenhouse vegetable industry and demonstrates that many farms have in fact experienced healthy growth in the last decade and can hardly be characterized as vulnerable. However, despite social differentiation among the Leamington greenhouse growers, all greenhouses rely heavily on offshore labour. It is not their vulnerability that forces them to import workers from abroad but their need to secure a reliable labour force ready to work on demand.

In part 2, I shift my focus from Ontario horticulture to the Mexican workers themselves, and in chapter 6 I relate the desire to work in Canada to the poverty that most workers experience before they join the program. I then explore the structural reasons that impel Mexican day workers to seek employment in Canada. Finally, I outline the mechanisms of their recruitment and discuss the characteristics of Mexican migrants selected to work in Canada. In Chapters 7 and 8 I explain why Mexicans seasonal workers accept their conditions of unfreedom. I point out in chapter 7 that Mexican labour is not necessarily cheap but that it is reliable in the sense that the workers are available for work on demand. I argue that the control over their recruitment exercised by the Canadian state and assisted by the Ministry of Labour and Social Planning and by the Mexican consulates, with full collaboration on the part of the growers, plays a pivotal role in shaping Mexican workers' acceptance of their conditions of unfreedom. Furthermore, Mexicans are available to work whenever there is a need because they have no social commitments in Canada to take them away from their jobs. Their compliance is also reinforced through paternalistic relationships between Mexican workers and their *patrones*. Finally, chapter 8 complements the argument made in chapter 7. It illustrates the ways in which Canadian-earned money is spent in Mexico, pointing out that very little of it is invested productively and that most of it is spent on consumer goods and children's education. This pattern of investment of remittances by Mexican seasonal workers makes it necessary for them to continue coming to Canada year after year in order to put all their children interested in education through school and in order to maintain the living standards to which they have grown accustomed. Therefore, they cannot risk losing their place in the program by challenging their conditions of unfreedom.

This book is based on research carried out in four stages. First begun in the Leamington area in 1996 and 1997, it moved to the village of San Cristóbal, in Guanajuato, Mexico and then returned to the Leamington area, focusing this time on greenhouse growers instead of on the Mexican workers employed by them. The final stage involved a survey of 311 Mexican participants in the so-called offshore program or of their spouses in eleven villages in the provinces of Guanajuato and Tlaxcala, in Mexico.

Finally, a short comment on the writing style adopted in this manuscript is in order. The opening pages of chapters 6, 7, and 8 are written in an informal style that is in sharp contrast to the rest of the book. There is a reason for this indulgence. Unlike the previous chapters, which deal with Canadian agriculture, the Leamington area, and Leamington greenhouse growers, chapters 6, 7, and 8 are based on my research on Mexican workers. Interviews with greenhouse growers were taped, and I was therefore able to use quotes from these interviews in chapter 3 and elsewhere in the book. When I first started conducting interviews with Mexican workers, I also used a tape recorder, but I soon realized that the workers were more likely to talk to me candidly if I did not use it. As a result, I do not have many quotes from Mexican workers in this book. To compensate for this deficiency, I have created short summaries of their stories, and I have also used composites at the beginning of each of the three chapters. It was my intention to write these composites in the way these stories would have been narrated to me if I had recorded them. I believe they embellish the book by giving a voice to my otherwise silent Mexican workers.

- DISCUSSES ECON. RESTRUCTURING IN INDUSTRY, AS IF (WE COULD EXTRAPOLATE FROM) THIS PROCESS ~~WERE ABLE~~ TO AG. IN CDA.! → LIT. REVIEW BARKS UP THE WRONG TREE! THE LIT. IS SO OLD THAT IT MISSES THE 1990s EXPANSION AND DOES NOT ACCOUNT FOR ▽ IN UNEMPLOYMENT.
- AFTER CHOOSING THE WRONG LIT. FOR REVIEW, SHE CRITIQUES IT ~~FOR~~ BECAUSE IT "CANNOT EXPLAIN THE CONDITIONS THAT MAKE THE EMPLOYMENT OF FOREIGN LABOUR STRUCTURALLY NECESSARY (P. 14).
- FROM HERE ON, SHE GOES ON TO DISCUSS HER MAIN CONCEPT/ OF "STRUCTURAL NECESSITY", MIGRANT WKR, UNFREE LABOUR, ETC. BASED ON OTHER LITERATURE. BUT BECAUSE THE PREVIOUS LIT. HAD DIFFERENT RESEARCH GOALS THAN HERS, BASOK'S CRITIQUE IS UNFAIR AND UNFOUNDED.
- YET, IT DOES SEEM LIKE HER ANALYSIS IS SIGNIFICANT IN THAT IT OFFERS THE EMPIRICAL BASIS TO ASSESS THE CDN. GUEST WKR. PROGRAM.
- WHAT SEEMS TO BE COMPLETELY ABSENT FROM THE BOOK OUTLINE IS A FRONTAL CRITIQUE OF CDN. POLICY.

PART ONE
Canadian Growers

2 The Farmers' Affliction

The Commonwealth Caribbean and Mexican Seasonal Agricultural Workers Program was put in place in 1966 in order to alleviate labour shortages experienced by Canadian fruit and vegetable farmers. Before World War II farming was predominantly a family business. Farm operators, assisted by their wives and children, provided most of the required labour, while nonfamily wage workers were hired occasionally. Since the 1940s, however, farms have been consolidating, and, at the same time, growers' families have become much smaller. Consequently, the demand for wage labour, hired predominantly on a seasonal basis, has increased. This labour has not been easily available to Canadian farmers, and several public and private initiatives to recruit farm labour have been undertaken. Unfortunately, these efforts did not guarantee a supply of reliable labour to the growers, and in the late 1950s they pleaded with the Canadian authorities to allow them to import workers from the Caribbean. Their pleas were answered in 1966 when the offshore labour program was put in place. It did not satisfy the total demand for seasonal workers, however, and growers, using the assistance of private contractors, brought Mexican Mennonite families and Portuguese migrants to work on their farms. The working and living conditions of these workers were deplorable, and their plight came to the attention of a special task force established by the

Department of Manpower and Immigration. To regulate their

working and living conditions, the task force recommended that an agreement similar to the agreements with Caribbean countries should be made with Mexico and Portugal, and the Mexican Seasonal Agricultural Workers Program was consequently put in place.

WAGE LABOUR DEMANDS IN CANADIAN AGRICULTURE

Until the 1940s farmers and their families performed most of the work on their farms. Since the 1940s, however, the proportion of hired workers in relation to self-employed farm operators and unpaid members of their households has increased significantly, as a result of farm commercialization and consolidation (Smit, Johnston, and Morse 1984, 6–11; Dawson and Freshwater 1975; Mitchell 1975, 26; Parr 1985, 102). The farm consolidation trend in Canadian agriculture is related to certain problems faced by family farms. The much faster rising cost of the implements required to produce agricultural commodities in relation to the price growers could charge for their produce, created a tremendous "cost-price squeeze" for many producers, who were forced to abandon farming and move to industrial jobs. On the other hand, other farmers borrowed funds to expand their land holdings and purchase machinery. This concentration of land and capital in the hands of larger producers resulted in an increase in hired wage-labour, both permanent and seasonal, despite mechanization (Mitchell 1975, 18–26; Ghorayshi 1987; Wall 1994).

The trend towards corporate consolidation in the agro-food industry continued in the 1980s and 1990s (Winson 1990, 1992; Shields 1992, 248) resulting in a greater reliance by farmers on wage labour. While mechanization did offset demands for labour in some sectors of agricultural production, it could not be a panacea for all farm labour shortages, for two reasons. First, not all crops can be harvested mechanically. And second, as Smit, Johnston, and Morse (1984, 8) suggest, some farmers did not wish to commit themselves to labour-saving technology, because of its elevated prices and the high interest rates charged by banks on loans required to purchase new machinery.

Whether labour-saving technology can be used in agriculture depends largely on the crop. Crops such as apples, tomatoes, cucumbers, and peppers depend heavily on manual labour. Tomato growers, for example, are heavy users of seasonal labour, despite their use of mechanical harvesters, since most operations are performed manually during planting and cultivating (Winson 1996, 104). Furthermore, since tomato processors manufacture juice and since juice tomatoes cannot withstand mechanical harvesting, manual labour continues to be a valuable commodity in this sector. Small nonmechanized farms relying on manual labour usually supply the required tomatoes to the processors (Winson 1990, 388).

More generally, the production of fruits, vegetables, and tobacco involves the most labour intensive agricultural activities, and they are therefore the most dependent on hired labour (Shields 1992, 249). Even though in Ontario the capital intensity of farms in the fruit and vegetable sector is above the provincial mean, most growers rely heavily on seasonal labour, many for three months or more each year (Winson 1996, 97–9). Flue-cured tobacco farming, in particular, is one of the most labour intensive forms of agriculture practised in Southwestern Ontario (Smit, Johnston, and Morse 1984, 26), and the Ontario flue-cured tobacco industry makes a significant economic contribution (22). In 1980, for instance, it occupied only 1.2 percent of Ontario's cropland, yet it accounted for 7 percent of the total value of all agricultural production in the province. Most of the labour that is required in this area is seasonal, because of the growth cycles of the crops (1).

Within Ontario, the areas that specialize in fruit, vegetable, and tobacco production and that rely on seasonal labour are along the north shore of Lake Erie, at the west end of Lake Ontario, and within the major tobacco growing region of the Norfolk Sand Plain. Others areas, such as York and Simcoe Counties and the north shore of Lake Ontario, particularly Prince Edward county, are also associated with the concentration of seasonal farm employment (Smit, Johnston, and Morse 1984, 11). While labour requirements in Ontario agriculture remain high despite mechanization, as mentioned, traditional sources of labour supply, namely farm operators and members of their households, have been drying up. On the one hand, the families of most farm operators have become considerably smaller, and on the other, employment opportunities elsewhere have drawn members of the household away from the farm. At the same time, the number of students employed in agriculture has gone down (Dawson and Freshwater 1975, 5). Consequently, Ontario has experienced a significant shortage of seasonal farm workers.

PUBLIC AND PRIVATE LABOUR RECRUITMENT INITIATIVES

Several public and private agencies have taken initiatives to provide needed seasonal labour to Canadian farms. One of the first initiatives dates back to 1896, when private railways supplied seasonal workers from Eastern Canada to the prairies. The supply of seasonal labour was particularly low during World War I and World War II. To help farmers, young men from urban centres were recruited for agricultural work by government representatives. In Ontario the Farm Service Force, formed in 1941, took charge of recruiting farm labour among children, youths, and adult day-by-day workers. In 1943 the Dominion Provincial Farm Programme was set up to recruit Canadian workers nationally. Additionally, the federal government supplied German prisoners of war, Japanese Canadian internees, and conscientious objectors, predominantly Doukhobors and Mennonites, to farmers. After the war labour shortages persisted, and some initiatives were taken to recruit among children, women, urban workers, and aboriginal people. In 1974 the federal government attempted to regulate the supply of farm labour by forming the Farm Labour Pool System (Smit, Johnston, and Morse 1984, 19–20; Satzewich 1991, 70–7; Dawson and Freshwater 1975, 14; Laliberte and Satzewich 1999). Under the Farm Labour Pool System workers from Quebec and the Maritime provinces were paid the bus fare to deliver them to the fields in Southwestern Ontario. They were trained for twenty-four hours and offered an for allowance room and board (Parr 1985, 102; Winson 1996, 100; CEIC 1980). Additionally, the Agriculture for Young Canadians program was designed to attract young Canadians eighteen years of age or younger to farm work during the summer vacation period and to future careers in agriculture (Dawson and Freshwater 1975, 14).

When internal sources of labour proved insufficient, the federal government turned to international sources. After World War II Polish war veterans and displaced people were contracted in war camps to work on Canadian farms (Haythorne 1960, 79–80). But most of them left their farms when their contracts expired, and some even did so before the termination of their contracts (Satzewich 1991, 85–98). Additionally, Canadian immigration authorities attempted to select immigrants suitable and willing to work on farms from

European applicants, but many of them either became farmers themselves or left farm jobs for more lucrative urban employment (Satzewich 1991, 98–107; Parr 1985, 103). Two initiatives addressed the specific needs of the tobacco industry. First, U.S. tobacco workers from the southern states were recruited to work in Canada and some of them may have settled in Canada permanently (Satzewich 1991, 108–10). Second, the International Youth Employment Exchange Program addressed the labour shortage by supplying several hundred foreign students to Canada each year to work on tobacco farms (Smit, Johnston, and Morse 1984, 20).

Despite these efforts Canadian tobacco, fruit, and vegetable growers continued to experience problems with labour shortages, mainly because of the high turnover rate among the workers they hired (16). Given the perishability of the crop, the high turnover jeopardised the very existence of family farms and made it imperative for them to secure a more reliable labour force.

GROWERS' LOBBYING: THE CARIBBEAN PROGRAM

In the years preceding the introduction of the Commonwealth Caribbean Seasonal Agricultural Workers Program, Ontario farmers experienced serious problems with labour shortages related to the economic expansion in other parts of the Canadian economy. First, this expansion reduced the unemployment rate, making it more difficult for farmers to recruit labour. And second, industrial expansion in some areas attracted rural people to better-paid industrial jobs, leaving farms seriously understaffed. Areas such as the Bradford Marsh, north of Toronto, Southwestern Ontario's tobacco belt, the Niagara Peninsula, and the counties of Kent, Essex, and Lambton experienced farm labour shortages in the 1960s as a result of better wages offered by industrial plants in Hamilton, St Catherines, London, Kitchener, and Windsor (Bezaire 1965, 15–17; Ward 1974, 18).

Consequently, Canadian farmers pleaded with the Canadian government to allow legal employment of Caribbean seasonal workers. But for several years the Canadian Department of Citizenship and Immigration and the Department of Labour remained reluctant to sponsor the movement of these "undesirable" migrants (Satzewich 1991). Canadian authorities insisted that the shortage of labour was a temporary problem and that it could be resolved if the growers would offer better working and living conditions to their workers. In response to the Ontario Fruit and Vegetable Growers Association's request for Caribbean workers, the chief of the Settlement Division of the Department of Citizenship and Immigration suggested that labour shortages were the result of "the almost complete lack of accommodation provided by the employers; the reluctance of growers to provide transportation; instability of wages, and the lack of arrangements to assure continuity of employment from one grower to another" (cited in Satzewich 1991, 158). Even when the Department of Citizenship and Immigration became sympathetic to growers' concerns, the Department of Labour persisted in its belief that agricultural labour could be recruited from internal sources and that growers were responsible for the labour shortage problems they faced (161–8). In response to the Essex County Associated Growers, the minister of labour stated:

> Your representation concerning the temporary admission of workers from abroad is based on the assumption that workers in the numbers required cannot be recruited in Canada. This, however, has not been established and it is felt that the requirements of agriculture can be met through a vigorous recruitment program involving local recruitment, day-haul movements, and the transfer of workers within and between provinces. It is the view of the government that we cannot import temporary workers at a time when the government is spending large sums to rehabilitate unemployed agricultural workers ... in other parts of Canada, when we are proposing to move workers who need employment from designated areas at public expense and when substantial sums are being spent through retraining and in other ways to move unemployed workers into employment in Canada ... If the growers were to offer wages and conditions to the extent proposed in respect to workers from Jamaica, taking into account the total cost of such a movement to the growers, it is felt that we can meet your labour requirements from within Canada. (cited in Satzewich 1991, 164)

Growers continued pleading with the Canadian authorities to recruit skilled agricultural workers from outside Canada, criticizing the National Employment Service for its failure to recruit reliable and experienced Canadian workers. Written on behalf of Essex County Associated Growers, the report by Ernest Bezaire that was published in 1965 reflects the opinion of many Ontario growers concerning the workers recruited by the National Employment Service from among the urban unemployed and Native people:

In many cases, the workers who came down had never been on a farm before. They were not physically able to perform the tasks required to stoop labour. They lacked experience and dexterity. Placed alongside an average farm worker, on a piece-work basis, the city workers managed to pick half the tomatoes the experienced worker picked and couldn't understand why it was that they only made half as much money.

Some of the workers who came down from Ottawa didn't last a whole day and threw up the job. Few lasted more than three days. Others shopped around through the neighbourhood to see where they could get a better deal.

The Indians were steady workers when they were working. But they were slow and were not too dependable after pay-day. Living accommodation was provided for these people in tourist cabins. Where the Indians went to bed with their clothes and boots on, it was invariably difficult to secure any such accommodation for them after the second week.

Very little of this imported urban labour was productive. On the average the productivity of labour, and this was proven wherever the labour was employed on a piece-work basis, was about half what would be expected normally.

Some of the Indians left half-way through the season to plant trees or to work as guides during the hunting season, indicating a preference for work closer to their homes. (9–10)

The report goes on to claim that among workers recruited in the cities there is a high preponderance of "drunks, half drunks, and winos who only work enough to get the price of a couple bottles of wine" (10). Bezaire attributes an increase in petty crimes, assaults, thefts, and the passing of bogus checks to the employment of urban workers in this area and cites one farmer who decided to put up his equipment for sale as saying, "I've had a guy pull a knife, I've had to stop men beating up women and I've seen elderly people over-come by heat. I've come to the conclusion that I can spare myself such experiences" (11). He recommends that "a limited number of skilled agricultural workers be recruited from outside Canada" and that two plane loads of these workers be sent, one to Essex County and one to the Niagara Peninsula (21). Similar arguments were made by various growers' lobby groups. Under the constant pressure on the part of Canadian growers, one of whom, Eugene Whelan, was a Liberal member of parliament for Essex South and a future minister of agriculture, the Department of Labour finally consented to the importation of Caribbean farm workers in 1966.

EXPANDING THE PROGRAM

In the early 1970s, in addition to the workers contracted through the Caribbean Seasonal Agricultural Program, two other groups of migrants were employed as seasonal workers in Ontario agriculture – Mexican Mennonites and Portuguese workers. Mexican Mennonites were brought to Ontario by brokers who charged a fee for recruiting Mexican families for seasonal farm work. They were put on a bus in El Paso, Texas, and carried to work in Ontario. They came with their entire families consisting of some twelve to fifteen members. Virtually all but the youngest children worked, while only fathers and possibly mothers were paid (Department of Manpower and Immigration 1973, 18–19). Portuguese adult males were recruited primarily in the Azores by a broker for a fee of five hundred dollars (20).

A special task force established by the Department of Manpower and Immigration examined working and living conditions of these (and other) farm workers. Its report demonstrated that both groups were over-exploited and lived under deplorable conditions, often in dirty barns. The following quote describes some:

> We saw another ramshackle building in which a Mexican family resides and for which they pay $30.00 monthly in rent. We were taken through the "house" and found a scene of almost indescribable squalor. Clothes were scattered over the floor and the three beds (there are nine in this family), there was no food in evidence and no cupboards where it could have been stored. A stove pipe led from the stove through the roof. Wrapped around the pipe at the roof intersection were lawyers of cardboard and newspapers to keep the rain out, for the reasons just mentioned and the fact that there is only one door. The family had just recently come to the attention of a minister and of the local health authorities due to the fact that the mother had the week previously been admitted to the hospital for childbirth by caesarian section. The baby girl

was still in hospital, but the mother was back in the fields the following week, i.e., two weeks after her operation. We were informed that *every one of the seven children, ranging in age from 8 down to 2 years, has at least one hernia*. Furthermore, they were ruptured when they came to Canada (last spring, illegally), and would continue to work despite this disability. (Department of Manpower and Immigration 1973, 35; italics in the original)

The task force called for agricultural agreements with Mexico and Portugal similar to the one that existed with various Caribbean countries and guaranteeing adequate wages, humane treatment of workers, livable accommodation, and transportation assistance (Sanderson 1974, 406). In 1974 an agreement between Canada and Mexico was signed that established the Commonwealth Caribbean and Mexican Seasonal Agricultural Workers Program. It forces private brokers out of business and regulates the living and working conditions of migrant workers.

MEXICAN SEASONAL AGRICULTURAL WORKERS IN CANADA

In 1974 only 203 Mexican men participated in the Commonwealth Caribbean and Mexican Seasonal Agricultural Workers Program. Since then, the number of Mexican participants has gradually increased, both in absolute numbers and in relation to the Caribbean inflow. While in 1987 Mexicans constituted only 22 percent of the arrivals, by 1998 their share of the total inflow of seasonal farm workers had almost doubled. While Mexican workers constitute the vast majority of the seasonal offshore workers in green houses, West Indian workers tend to be hired predominantly to harvest tobacco and apples (Verduzco 2000, 340). It is possible that Canadian growers regard the inability of Mexican workers to speak English as an advantage, since they are less likely to talk back to their employers and demand improvements in their working and housing conditions. Furthermore, there is virtually nothing to distract them from their total commitment to work . Whereas West Indian workers can easily communicate with Canadian residents in English and have made friends with some of them (Knowles 1997, 100), Mexican workers, who cannot do so, are less likely to socialize with people off the farm. As will be discussed in chapter 7, growers prefer their workers

Table 2.1
Seasonal Workers Employed in Canada

Year	From Mexico		From the Commonwealth Caribbean[1]		Total
	Percentage	*Number*	*Percentage*	*Number*	*Number*
1987	22	1,347	78	4,777	6,124
1988	26	2,092	74	5,955	8,047
1989	31	3,504	69	7,798	11,302
1990	36	4,168	64	7,410	11,578
1991	37	4,102	63	6,984	11,086
1992	38	4,216	62	6,880	11,096
1993	38	3,878	62	6,328	10,206
1994	39	3,870	61	6,054	9,924
1995	37	3,799	63	6,469	10,268
1996	40	4,177	60	6,265	10,442
1997	40	4,537	60	6,805	11,342
1998	43	5,233	57	6,937	12,170

Source: Statistical reports compiled by FARMS.
[1] Includes Barbados, Eastern Caribbean, Jamaica, and Trinidad and Tobago.

not to make friends with outsiders. And because many of the growers in the Leamington area speak Italian, which is similar to Spanish, the inability of the Mexican workers to speak English does not bother them. Mexicans learn to understand some Italian spoken by the growers, and vice versa. The absolute number of Mexican seasonal workers employed in Canada in a given year increased almost four times from 1987 to 1998 (table 2.1). In the last few years, slightly more than five thousand Mexicans have been able to participate in this program annually. On average they are employed between eighteen and nineteen weeks per year.

Most participants in the program are men. In 1993 women constituted only about 2 percent of the participants. In the years that followed that figure went down to 1 percent, although in 1998 it increased to 2.7 percent. The percentage of women is even lower among the Caribbean participants in the program (table 2.2).

In greenhouses and in the field Mexican seasonal workers harvest tobacco, ginseng, apples, peaches, strawberries, corn, tomatoes, peppers, cucumbers, lettuce, asparagus, broccoli, carrots, cabbage, turnips, potatoes, beets, onions, squash, zucchini, and cauliflower. They work, as well, in canning plants and nurseries in rural communities in Ontario, Quebec, Alberta, and Manitoba. In 1998 more

Table 2.2
Female Workers, Mexican and Caribbean Seasonal Agricultural Workers Program

	Mexico		*Caribbean*		
Year	*Total*	*Percentage*	*Total*	*Percentage*	*Total*
1993	77	71	32	29	109
1994	48	62	29	38	77
1995	53	63	31	37	84
1996	57	69	26	31	83
1997	67	71	27	29	94
1998	141	85	25	15	166

Source: FARMS.

Table 2.3
Employment of Seasonal Workers, Various Crops, 1997 and 1998

Crop	*Number of Workers, 1997*	*Number of Workers, 1998*	*Percentage Change*	*Number of Employers, 1997*	*Number of Employers, 1998*	*Average per Employer, 1998*
Fruit	2,065	2,281	10.5	254	246	9.3
Vegetables	2,145	2,383	11.1	302	306	7.8
Tobacco	4,382	4,892	11.6	611	654	7.5
Greenhouse	838	1,109	32.3	142	146	7.6
Nursery	293	342	16.7	31	36	9.5
Canning	541	343	–36.6	14	11	31.2
Apples	1,792	1,461	–18.5	130	127	11.5
Ginseng	309	339	9.7	51	57	5.9
Total	12,365	13,150	6.0	1,461	1,514	8.7

Source: FARMS.
Note: Seventy-four employers had multiple crops in 1997; 69 employers had multiple crops in 1998. Totals are higher than in Table 1.1 because they are based on a sum of "arrivals" and "transfers."

than one-third of seasonal workers were employed in the tobacco industry. While the greenhouse sector employed only 8 percent of the workers, its foreign seasonal labour force has experienced the most significant growth in the last few years, which compensates for the decline in the number of workers employed in the canning, food processing, and apple industries. The average number of foreign seasonal workers per employer varied from 5.9 for the ginseng crop to 31.2 in canning and food processing, with an overall average of 8.7 workers per employer in 1998 (table 2.3). Eighty percent

Table 2.4
Communities with Foreign Seasonal Workers, 1998

Community	*Number of Positions Approved*
Barrie	136
Brockville	16
Chatham	374
Cobourg	232
Collingwood	944
Cornwall	30
Goderich	191
Guelph	11
Hamilton	316
Kitchener	10
Leamington	1,398
Lindsay	13
London	382
Mississauga	200
Newmarket	466
Oshawa	174
Ottawa	30
Picton	31
Sarnia	46
Simcoe	4,680
Smiths Falls	4
St Catharines	1,423
St Thomas	350
Stratford	12
Tillsonburg	1,536
Trenton	62
Wallaceburg	148
Woodstock	176
Total	13,391

Source: FARMS.
Other communities that have had foreign seasonal workers in previous years are Aylmer, Brampton, Brantford, Burlington, Exeter, Hawkesbury, Listowel, Milton, Oakville, Orillia, Owen Sound, Port Colborne, Stoney Creek, Welland, and York.

of these workers are sent to work in over forty communities in Ontario where farm labour shortages are most severe. Altogether five communities – Simcoe, Tillsonburg, St Catharines, Leamington, and Collingwood – receive 9.98, or almost three-quarters of all foreign migrant workers (table 2.4). Simcoe alone provides work for 4,680 (35 percent) of the participants in the Commonwealth Caribbean and Mexican Seasonal Agricultural Workers Program. The other four communities receive between 7 and 11 percent of the program participants each. Leamington provides employment for 10 percent of seasonal farm workers, and most of the workers sent to that area are Mexican.

ADMINISTRATION OF THE MEXICAN SEASONAL AGRICULTURAL WORKERS PROGRAM

The Commonwealth Caribbean and Mexican Seasonal Agricultural Workers Program is administered in accordance with the following general principles and operational guidelines. During peak periods offshore seasonal agricultural workers may be authorized for employment in commodity sectors where the supply of Canadian workers is deemed insufficient. Employers are requested to formally advise the local Human Resource Centres of Canada (HRCC) office of their total agricultural labour requirements at least eight weeks in advance. In its turn, HRCC assesses each employer's request on the basis of the adequacy of the job offer, the availability of qualified Canadian farm workers, and the employer's previous experience in attracting and retaining workers. Requests from employers to "name" workers and/or to receive up to the same number of foreign workers as was previously authorized the year before is approved by HRCC on a priority basis. Employers are required to pay a user fee to Foreign Agricultural Resource Management Services (FARMS), which is appointed by Human Resources Development Canada to administer the program. FARMS, which was incorporated federally in 1987, is a grower organization funded exclusively through a user fee. In the province of Ontario it consists of representatives of the following commodity groups: the Ontario Food Processors Association, the Ontario Fruit and Vegetable Growers Association, the Ontario Greenhouse Vegetable Producers Marketing Board, the Ontario Vegetable Growers' Marketing Board, the Ontario Nurseries Trades Association (Landscape Ontario), the Ontario Flue Cured Tobacco Growers Marketing Board, the Canadian Nurseries Trade Association, and the Ginseng Growers Association of Canada. FARMS is managed by officials of Human Resources Development Canada (HRDC).

Terms and conditions of employment have been negotiated by the government of Canada, the participating foreign governments, and representatives of the Canadian Horticultural Council. The

Caribbean and Mexican workers are to be provided with free accommodation approved by the Ministry of Health and receive the prevailing rates of pay for the work performed. The employers are expected to pay the travel agent the cost for their workers of two-way air transportation from Mexico City to Canada. They are also to make arrangements to transport the workers from their point of disembarkation in Canada to the place of employment and, upon termination of the contract, back to their place of departure from Canada. These expenses are partly reimbursed to the employer through payroll deductions from the workers' salary. The reimbursement is calculated at a rate of 4 percent of the workers' gross pay, and the aggregate payment is not to be less than $150 or greater than $425 per employee. The employers also make other deductions from workers' salaries: Canada Pension Plan contributions, Employment Insurance, Voyageur Insurance premiums, and workers' compensation insurance premiums. Income tax is withheld from workers whose income exceeds a certain minimum. In 1997, for instance, it was withheld from married workers whose earnings exceeded $11,836 and from single workers whose income exceeded $6,456 (increased in 1999 to $12,836 and $6,956, respectively).

The Agreement for the Employment in Canada of Seasonal Agricultural Workers from Mexico stipulates that workers are not to be moved to another area of employment or transferred or loaned to another employer without the consent of the worker and the written approval of HRDC and the government agent. They are authorized only to perform agricultural labour for the employer to whom they are assigned. Employment of foreign workers by unauthorized employers or for nonagricultural work is subject to a $5,000 fine or two years imprisonment, or both.

The workers agree to work and reside at the place of employment or at any other place provided by the employer and approved by the government agent. They are to perform the duties of the agricultural work assigned to them and obey and comply with all rules set down by the employer relating to safety, discipline, and the care and maintenance of property. More specifically, they are to keep the living quarters provided to them by the employer in a clean condition. They are not to work for any other person without the approval of HRDC, and they are to return promptly to Mexico upon the termination of their employment authorization visa (FARMS 1999, 21–5).

On the Mexican side, the program is administered by the following agencies: the Ministry of External Relations (through the Mexican Embassy and general consulates in Toronto, Montreal, and Vancouver), the Ministry of Interior (through the National Migration Department), the Ministry of Labour and Social Planning, and the Ministry of Health. Most of the work is carried out by the Ministry of Labour, which is in charge of the selection of the workers, the Ministry of External Relations, which is responsible for the documentation of the migrant workers, and the Mexican consulates in Canada. The mandate of the consulates is to

- ensure respect for the working and living conditions specified in the memorandum of understanding between Mexico and Canada in regard to the Seasonal Agricultural Workers Program, the operative norms of the memorandum, and the Agreement for the Employment in Canada of Seasonal Agricultural Workers from Mexico;
- receive agricultural workers at the airport in Toronto, Montreal, and Vancouver, orient them in regard to their rights and obligations, and present them to the representatives of the farms;
- visit the farms in order to inspect the living and working conditions of the agricultural workers;
- respond to workers' requests and participate in the resolution of problems emerging between workers and their employers;
- inform the Ministry of External Relations and Canadian authorities with respect to the arrival, transfer, and return of workers, as well as with respect to any break-ups of contracts that may occur; and
- propose to the Ministry of External Relations strategies and actions to be taken in order to negotiate better conditions for Mexican workers.

EMPLOYMENT CONDITIONS OF MEXICAN SEASONAL WORKERS: THE OFFICIAL STORY

The terms and conditions of employment of foreign workers are specified in an employment agreement to be signed by each employer and worker and by the government agent. This agreement states that employment should be for not less than 240 hours for a period of six weeks or less and that it should not exceed eight months. Additionally, the employment should occur during the peak season, as identified by the regional agricultural program coordinator for the approved commodity sectors. In the case of a transferred worker, employment is to consist of a cumulative term of not less than 240 hours.

The Agreement for the Employment in Canada of Seasonal Agricultural Workers from Mexico also stipulates the following. The normal working day is not to exceed eight hours, but the employer may request the worker to extend his or her hours when there is an urgent need. For each period of six consecutive days of work, the worker is entitled to one day of rest, but when there is an urgent need, the employer may ask the worker to delay that day until a mutually agreeable date. The employer is to give the worker a trial period of fourteen working days during which the worker cannot be discharged except for the sufficient cause of a refusal to work during that trial period.

The employer also agrees to provide free approved accommodation to workers and reasonable and proper meals. If the workers choose to prepare their own meals, employers are required to provide cooking utensils, fuel, and facilities without cost to the workers and to allow for a minimum of thirty minutes for meal breaks. The employer agrees to allow HRDC access to all information and records necessary to ensure contract compliance. The workers are to receive weekly wages calculated as the greatest of

- the minimum wage for workers provided by by provincial legislature,
- the rate determined annually by HRDC to be the prevailing wage rate for the type of agricultural work being carried out, and
- the rate being paid by the employer to Canadian workers performing the same type of agricultural work.

If no work is possible, for whatever reason, the workers shall receive a reasonable advance to cover their personal expenses (FARMS 1999, 22).

SUMMARY

While demands for wage labour have been increasing in Canadian agriculture and especially in the tobacco and fruit and vegetable sectors, farm operators have had a hard time recruiting workers. Various public and private initiatives to recruit labour within Canada or from among new Canadian immigrants have not resolved the problem of the high turnover rate among workers in these sectors. In the late 1950s Canadian growers appealed to the Canadian government to allow Caribbean migrants temporary visas to work during the months of high labour demand. Despite their initial reluctance, the Department of Citizenship and Immigration and the Department of Labour authorized the Caribbean offshore program in 1966. West Indian workers did not fill all the gaps, however, and Canadian farmers continued recruiting other foreign workers, namely, Portuguese and Mexican Mennonites, using private contractors. The extremely abusive working and living conditions of these workers came to the attention of a special task force created by the Department of Manpower and Employment, which recommended the extension of the Caribbean Agricultural Seasonal Workers Program to Mexico and Portugal. While no agreement has been signed with Portugal, Mexico joined the program in 1974. The program, based on formal intergovernmental agreements and administered by several governmental and private agencies, guarantees a regular supply of labour to Canadian farmers and regulates working and living conditions of foreign agricultural workers.

THE EMPIRICAL DATA AND ANALYSIS OFFERED BY BASOK IS THE STRONGEST OF HER BOOK'S CONTRIBUTIONS. SHE DOES A FINE JOB OF CONTEXTUALIZING HER LEAMINGTON CASE STUDY WITHIN THE OVERALL CANADIAN AND ONTARIO CONTEXTS. THIS COMMUNITY ALONE RECEIVES 10 PERCENT OF ALL MIGRANT WORKERS FROM THE CCMSAWP, AND ONTARIO GETS 80% OF THE CANADIAN TOTAL.

BY THE LATE 1950S, CDN FAMILY FARMERS HAD BEEN FACING SEVERE PROBLEMS OF INCREASING INPUT COSTS AND DECLINING PRICES FOR THEIR PRODUCTS. THIS "COST-PRICE" SQUEEZE LED TO THE TYPICAL PROCESS OF CONCENTRATION AND CENTRALIZATION IN FARMING, WITH THE CONCOMITANT INCREASE IN HIRED LABOUR. THE LATTER RESULTED BECAUSE NOT ALL FARMING PROCESSES MAY BE MECHANIZED, GIVEN ALSO THE HIGH FINANCIAL COSTS OF DOING SO. THE POST-WAR CDN ECON. EXPANSION HAD LED TO ∇ UNEMPL., AND △ RUR-URB. INDUSTRIAL WORK. => △ PROBLEMS FOR FARMERS TO FIND WORKERS. => INSTRUMENTAL STATE (P.32).

3 Labour Problems: The Leamington Story

Leamington is a rural community in Essex County some forty-five kilometres southwest of Windsor. Located in the south and in close proximity to the Great Lakes, Essex County has the earliest spring, warmest summer, and longest growing season in eastern Canada. Because of its location and because of the variety of productive soils found there, a wider range of crops has been grown in the area than elsewhere in Canada (Morrison 1954, 319). The favourable natural endowments have not been matched by a steady supply of the labour required to maintain high production levels, however. In order to appreciate the extent of the labour shortage problems that plague this community, it is useful to sketch the history of agricultural development in this region.

AGRICULTURE IN LEAMINGTON: AN HISTORIC JOURNEY

Specialized farming originated in the Leamington region in the 1880s. Peaches became the first market commodity, followed by tobacco, vegetables, and other fruit. Of the wide variety of fruit grown for market, grapes came to play an important role (Morrison 1954, 12, 150–3; Snell 1974, 90–1). Around the turn of the century, canneries and tobacco processors started production in the region, and by 1931 Leamington had become one of Ontario's leading cigarette manufacturing centres, competing with such areas as Simcoe, Delhi, Aylmer, and Tillsonburg (Snell 1974, 93; Sheik 1987, 28).

Agriculture experienced steady growth in the last few decades of the nineteenth century. In 1900 the *Toronto Globe* described Essex Country as "Eden without the serpent." The paper noted that "Market gardening along the Essex front has developed to a wonderful degree within the past year. In this line there is a large export trade to Detroit, which is eager for the Canadian products. Dozens of loads of garden stuff cross the river daily to be exchanged for silver dollars" (quoted in Morrison 1954, 193).

This growth continued into the twentieth century. At the beginning of the twentieth century two important developments gave a boost to the regional economy. The first was the establishment of a food-processing plant owned by the Heinz company in 1908. In 1910 the company began manufacturing ketchup, and the increased production of ketchup in Leamington turned the town into the Tomato Capital of Canada. In 1913 the company expanded its production of ketchup and canned tomatoes, and in 1930 it started producing tomato juice. Baby food varieties were added to the line of products in 1934 (Snell 1974, 97–101), and the company has been expanding ever since. Heinz is still the major food processor in Essex County, processing large amounts of tomatoes, cucumbers, and other fruits and vegetables.

The second important development was the opening of an experimental station in Harrow. First designed as a tobacco research centre, the Harrow station soon became a federal testing laboratory that developed hybrid corn, soybeans, grains, vegetables, and fruit and that introduced innovations into dairy and poultry farming. Local growers have benefited tremendously from the soil and seed tests performed at this research centre. These two developments also provided a major impetus to the greenhouse industry, which was introduced in the early 1940s (Snell 1974, 94). However, while the production of tomatoes and other vegetables and fruit has increased, the tobacco industry has declined. Many tobacco processing plants have moved elsewhere (95).

THE ETHNIC COMPOSITION OF LEAMINGTON

Until the middle of the twentieth century, settlers in the Leamington area were mainly of British descent, but in the late 1920s they were joined by Germans from Swabia (a historic region which currently includes parts of Germany, Austria, Hungary, and the Czech Republic), as well as Italians. The first Italians allowed to settle in Canada were required to pay a two-hundred-dollar bond with a provision that this money would be refunded if the men worked in agriculture for the first two years upon arrival in the new country. By first working as share-croppers, many were able to save enough money to eventually buy land and sponsor their wives and children to join them (Cornies 1977b, c).

The period after World War II witnessed the arrival of new waves of immigrants to Leamington, including Germans, Italians, and Portuguese. Most of these migrants also took farming jobs, eventually becoming land owners (Cornies 1977a, b, c). These early migrants were followed by others from a various of regions worldwide.

According to the 1996 population census, some 16,000 people of diverse ethnic backgrounds reside in Leamington. Even though some 4,800 stated in the census that they were immigrants, close to 5,400 people claimed that their mother tongue was neither English nor French. German and Portuguese speakers constitute the largest communities (2,165 and 1,065 people, respectively). In addition, 700 people reported Arabic and over 500 reported Italian as their mother tongue. Among the German-speakers very few were actually from Germany. Approximately one-half were German-speaking Mexican Mennonites. The rest of the German-speaking residents must have been other Mennonites born either in Russia, the United States, or Canada.* The Arabic speakers were predominantly from Lebanon. Other immigrants whose mother tongue was neither English nor French came from various regions, including Western and Eastern Europe, South Asia, South-East Asia, the Caribbean, the Middle East, West Africa, and Central America.

There is no available information on the ethnic origin of Leamington growers. However, by relying on the town of Leamington directory of greenhouse growers I was able to use the growers' surnames as a rough indication of their ethnic background. Of 75 growers listed in the directory, 31 had Italian surnames, 18 had German surnames, and the rest represented diverse ethnic origins,

* According to the 1996 census, 140 people stated they were from the Russian Federation, but only 15 claimed Russian as their mother tongue. It is possible that most were German-speaking Mennonites.

including Portuguese, Chinese, Yugoslavian, British, and French. Among those with German surnames, most were probably descendants of Mennonite settlers.

THE MENNONITE SETTLEMENT

The Leamington Mennonite community has been shaped by four waves of immigration. The first thirty Mennonite families who came to live in Essex County were part of the 1920s refugee movement from Russia to Canada. Following the Russian Revolution, Mennonites experienced famines, epidemics, and political violence. The first contingent of refugees, who feared that their religious freedom would be jeopardised by the new atheist state, numbered one thousand people who originated in a Ukrainian colony of Molotschna. They moved to Waterloo, Ontario, where they were assisted by Old Mennonites, whose ancestors had moved to southern Ontario as part of the Loyalist influx from German colonies in Pennsylvania in 1776. Enticed by the prospects of employment in Essex County, some thirty families left Waterloo and came to settle in this region between 1924 and 1926 (Driedger 1972, 6–24; Kliewer 1997, 6–8; Snell 1974, 117).

The second wave of Mennonite immigration to the Leamington area was related to the internal movement of Russian Mennonites within Canada. From 1923 to 1930, a total of twenty thousand Mennonites left Russia for Canada. Most of them settled in Saskatchewan and Alberta (Driedger 1972, 14), while some of them joined the survivors and descendants of a previous wave of Russian Mennonite immigration to Canada who were residing in Manitoba. That wave came between 1874 and 1876, when from five to eight thousand Russian Mennonites had been invited by Canada to settle in Manitoba. They were eager to leave Russia because of the introduction of compulsory military service, impositions placed on religious and economic freedom of Mennonite colonies, and the shortage of available farmland in Russia. Canada, on the other hand, offered them land, greater freedom, and economic assistance (Driedger 1972, 73; Kliewer 1997, 7–8). Some Mennonites residing in the Western provinces eventually moved to Essex County as a result of having married Mennonites from this region. In addition, in the 1930s other Mennonite migrants from the prairies and Northern Ontario came to the Leamington area hoping to find better economic opportunities (Driedger 1972, 50, 75).

The third wave of immigration consisted of the post–World War II Russian Mennonites who had been taken to Germany by the retreating German army. About seven or eight thousand of these Russian Mennonites came to Canada between 1948 and 1951. They were among thirty-five thousand Russian Mennonites who lived in Germany as refugees, hiding from the Russian army for fear of being deported to Russia and sent to Siberia. Upon their arrival in Canada, they were placed in the care of the Red Cross, which made arrangements for them to be transported to various parts of Canada. Among them were relatives and friends of Leamington Mennonites, who invited them to join them in that area (Driedger 1972, 75–7). First employed in two brickyards – Broadwell and Jasperson – and on farms, they came to fill other jobs in the Leamington economy (Driedger 1972, 25–6). Many of their descendants eventually purchased land, and today many own farms and employ domestic and imported labour.

The fourth wave of Mennonite immigration to Essex County was comprised of Mexican Mennonites who were actually the descendants of immigrants from Canada to Mexico and who have now become an important part of the Leamington labour force. In 1922, at the invitation of President Alvaro Obregón, twenty thousand Mennonites came to Mexico from Canada to settle on 247,000 acres of land in the state of Chihuahua. The main reasons for the move were the scarcity of land in Canada and pressure exerted by the Canadian government on the autonomy of Mennonite communities. Despite its original promises, the Canadian government had begun to infringe on the Mennonites' religious freedom, forcing them to educate their children in English-language public schools. Furthermore, a law requiring all males between the ages of sixteen and sixty-five to register with the authorities had caused great concern among Mennonites, who feared that it was a preliminary to military conscription (Sawatzky 1971, 12–30; Janzen 1988). In exchange for the promise that their children would be forever exempted from the educational laws of Mexico and from serving in its armed forces, the Mennonites agreed to purchase land from the Mexican government. Mennonite settlers were also exempted from paying taxes for fifty years

(Sawatzky 1971, 36–46). Agriculture, particularly the dairy industry, became the backbone of the Mennonite economy in Mexico.

However, growing economic hardships, as well as internal changes and tensions within the Mennonite colony and church, impelled descendants of these migrants to return to Canada, starting in the 1950s. First coming as seasonal workers, most of them subsequently chose to settle in Canada. They have generally looked for areas with existing Mennonite communities and with a high demand for labour (Jansen 1988; Kliewer 1997, 10–11). Leamington is one such region. Among the Mennonite immigrants of the fourth wave, many already had Canadian citizenship, and others who had failed to obtain it while living in Mexico have been able to apply and receive it upon their immigration to Canada. Mexican Mennonites residing in Canada speak mostly Low German, although the men are also able to communicate in Spanish and to a certain degree in English. Language and history are the elements that link these Mennonite migrants to the earlier waves. However, whereas Mennonites who have lived in Mexico have, to a large degree, maintained their religious beliefs, practices, and traditional life-styles, Canadian-born Mennonites living in Leamington seem to have assimilated to the Canadian mainstream culture.

LABOUR SHORTAGES IN LEAMINGTON

This section is based predominantly on interviews with forty-five greenhouse growers, forty of whom produced vegetables and five of whom grew flowers or potted plants. Of the forty vegetable growers, twenty-four grew tomatoes, seventeen grew cucumbers, and five grew peppers. In addition to greenhouse vegetables, sixteen producers grew various field vegetables or fruits.

Like other Canadian farmers discussed in chapter 2, Leamington growers have experienced numerous problems recruiting and retaining reliable local workers, even when they raise wages. Those who do come to work lack discipline and commitment to their employers, and some eventually quit their jobs without notice. Most of them take time off work without consideration for the growers' needs. At the same time, Leamington-area growers have become more dependent on hired labour, both year-round and seasonal, than farmers in other parts of Ontario (tables 3.1 and 3.2).

Table 3.1
Percentage of Farms Reporting Paid Labour

Region	*Year-Round*	*Seasonal*
Ontario	16	31
Essex	17	33
Leamington area	25	44

Source: 1996 Agricultural Census.

Table 3.2
Average Number of Weeks of Paid Labour

Region	*Year-Round*	*Seasonal*
Ontario	103	37
Essex	182	72
Leamington area	308	126

Source: 1996 Agricultural Census.

Consequently, problems of labour shortages affect these growers more than other Ontario farmers.

In order to attract labour, Leamington growers advertise in local newspapers, submit their requests to local unemployment or welfare offices, and participate in national labour pool programs. In 1999 Essex County growers offered tours to a group of jobless Hungarian refugees from Toronto to attract them to the area to fill some of the estimated three hundred to five hundred job vacancies. While the Hungarians seemed interested, nothing resulted from this guided tour (Hill 1999b). Often no applicants call for several weeks after job adds are placed with employment agencies. One grower recalled that "We need[ed] to hire three people. So we told this to welfare and they said, 'We will supply you with three people.' And they came back and said, 'We can't get you anybody. Nobody wants to come out.'"

Stories like this are so common that Leamington growers have become extremely skeptical about the ability of the institutions and programs to supply farm labour. At the same time, the approval of growers' applications for offshore workers is contingent on demonstrated efforts to recruit domestic labour. The growers therefore persist in submitting their requests for farm labour to government employment services, despite the disappointing results.

As mentioned, very few domestic workers who accept farm jobs keep them. One grower said he "tried the locals and it didn't work. They quit. We teach them everything, and it looks like they were gonna stay. And then we were picking and in the afternoon they wouldn't come back. And the other locals I had they quit after five hours, four hours. Before I went with the offshore I tried forty people." As another complained, "A lot of people start, and two days later or a day later or an hour later they would say, 'I don't want to do this.'"

Leamington growers have attempted to hire workers from other Canadian provinces through the Farm Labour Pool System (discussed in chapter 2), but they have concluded that workers recruited through this program are no different from the local labour. One grower expressed his frustration this way: "I don't disagree with 'Canadians first.' But you get those Northerners, the Frenchmen, and when the geese would start flying they get homesick. And they want to go home. So how are you going to finish your crop?" Another grower reflected that "A few years ago we went through the program where we hired people from Quebec, from all over Canada. And I had one guy who stayed for a month-and-a-half, and he said, 'Oh, I've got to leave. I can't take it no more.' We went through five, six at one time and the one guy stayed the longest – a month – a month and a half. And the other guys – several days ... It's a waste of time."

The regional Social Services Department has directed some welfare assistance recipients to the farm jobs. The experience of Leamington growers with these workers has been mainly negative. One grower complained that "It's a joke. You take them into a greenhouse and you show them work. And they tell you, 'You better not trust me to do that because I'll break something.' So what do you do? They don't want to do it but they are made to come out here. And when you're made to come out here you say, 'Kiss my ass.'"

Many growers blame the social welfare system for making it too easy for people to stay at home and "drink beer," instead of securing employment. One Leamington grower commented that "As a Canadian, you know there is welfare. They come here. I had some here. They wouldn't work more than two days in a week after three weeks, and they will find excuses. 'Wife's sick in bed.' They come up with so many excuses. Because they know the system, because this way they won't be cut off from the welfare." Another grower reflected that

They don't want to work first of all ... It's getting more and more difficult to find people who are willing to work here and not who are put here to appease their welfare reps or unemployment people and the government to give them a hand-out. They just have to show some good will on their part and they know full well all they have to do is quit and say, "The work is too hard," or "I don't like it," or "I am sick," or "I am allergic", or something or other, and they'll go right back on welfare. They've put in their week or two of work. They've done their bit to show they're trying, and they are right back on it again. And that is all the time at our expense, because these people are supposed to be productive and we're not supposed to carry them for two weeks just to fulfil their agenda. That's not our job.

Another farmer observed that "They come out and you show them the work and they say, 'I'm not up to it' for various reasons, depending on the season. Then they say, 'Can you sign my forms so that I can take them back to the office?'" Another farmer remembered that "There was another individual, he came one day, and next day he said, 'What am I supposed to do?' 'Well do what you were doing yesterday with the other guys.' 'No,' he says, 'I am not going to do that.' I say, 'Why not?' And he says, 'It's too hard for me.' 'Well, that's the job right now, and when we're done we will move on to something else.' No, he just walked off the farm. And here you're depending on these people ... On this farm, it's hundreds of thousands dollars, and if you bring somebody in that doesn't care, it's devastating." Some workers referred by the social services department stay on the job but put no effort into it. As one grower recollected about one such worker, "He threw in his time, he didn't make no effort to work hard. And he told us about it, and we couldn't get rid of this guy. He said, 'They won't pay for my school if I don't stay here till August.'"

In September 1999 Ontario premier Mike Harris proposed to expand the province's work-for-welfare program to include farm labour. Under this program, able-bodied welfare recipients are required to seek training or work to keep receiving social assistance. The reaction on the part of the growers was swift. Nick Ingratta, chairman of the Ontario Greenhouse Vegetable Producers'

Marketing Board, commented: "I have a hard time envisioning people on welfare – being forced to work – being reliable. I have a hard time with that ... My fear is that once you tell people they have to work, the potential (in the greenhouse sector) is for more damage than good" (Schmidt 1999).

Growers have used various incentives to encourage their workers to stay on the job, but they still end up losing their workers. One grower mentioned how he had helped six Iraqi workers bring their families over. He had covered the expenses of the sponsoring the families, hoping to deduct these costs from the workers' pay later. But all these workers had left right after their families arrived. The last worker he had helped "didn't show up on Monday, which was a holiday (Victoria Day), and I told him, 'We are very busy.' And he is just not showing up. And he is the sixth person now. Which is a frustrating thing." Another farmer described "one guy here just recently. [He] really upset me. We were really, really slow, but I hired him. I told him, 'I'll give you the hours now; the work is not too hard.' I gave him a lot of hours, just meaningless jobs, just so that I would have him when the plants came. And he disappeared. And that hurts. But there are lots of stories like this."

Some growers even told stories about workers who cause damage to the crop in order to get fired and have their welfare assistance restored. One grower bitterly recollected that 'This guy had people who were on social assistance in Windsor. The guy said, 'I don't wanna be here.' And this guys says, 'Why not? You've got to work.' And to make the long story short the farmer showed the guy the job, worked with him for quite a while. As soon as the farmer left the guy went through the crop and broke off all the hearts [without which a plant like a tomato cannot survive] and cost him in the neighbourhood of fifteen thousand dollars of damage in one day. And the farmer has no recourse. You're supposed to take a chance of putting people like that." Another remembered "I've had people who said that they were sent by unemployment insurance, and they asked me 'Do you want me to go wreck a few plants or should I go home as it is?' ... That's the usual story."

Leamington growers felt that even higher wages did not entice local workers to keep the farm jobs. As one grower observed, "I can't get more than four months out of them. And I raise the wage all the time. I've had people stay a year before. Even if they get tired in a year, they just go to another place just for the same wage. I don't know what it is. I can't figure it out. It's just that they get tired and they move on. I see people just moving around at different farms with all the farmers I know ... They don't get any more money, they just move around." Another grower observed, "If I pay them a dollar more one way or another, if I know he is going to stay I will be more than happy to. But if they stay for two months and it takes me literally a month to train them ..." A similar comment was that, "Even if you pay the Canadian workers more, they won't stay. Even if you pay them $15, $10. It's the environment." Another grower recollected that "Before offshore workers come I need local people here. I go put my name on unemployment and hope that workers come. I had my name on unemployment for two weeks and no one showed up, and I called the lady again and she told me, 'Raise it up to $7.50 instead of $6.90.' So I raised it to $7.50, and in two weeks and a half one person showed up, and he worked for two hours. And I said nothing to him. He couldn't keep up with my workers, the offshores I had. And he said, 'I can't do this job.'" Another grower contended that "With local help, I really don't think it would make a difference, you pay them $10 or you pay them $8. Because if they don't like it, they'll just try to find a different job, a different farmer ... And you treat them real good, local people, because you want to keep them, but the grass is always greener somewhere else."

For Windsor residents, commuting to Leamington by bus can be extremely inconvenient. Because only one bus travels in each direction every day, arriving in Leamington at 9 A.M. and leaving for Windsor at 4 P.M., it is virtually impossible for anyone interested in a farm job to rely on buses. The $26 bus fare is also rather high for workers making close to $7 per hour.

In January 1999, in recognition of these difficulties, the local HRDC office launched a joint effort with the Ontario Greenhouse Vegetable Producers' Marketing Board to bus welfare recipients to Leamington. Growers paid seventy-five dollars per week to cover the workers' bus fare and were also responsible for picking the workers up at the bus station and driving them back. Some growers estimated that at least a thousand different people came to Leamington through this program in the first two months, but very few kept their jobs. Some returned to school a couple of months after the program began, but most quit because they did not like the working conditions in the greenhouses. One grower complained that the bus schedule made the program unacceptable because during the harvest season workers were expectes to put in twelve hours per day or more, but the bus for the program arrived in Leamington at 8 A.M. and departed at 5 P.M. Although the program cost the growers a great deal, it failed to solve their labour shortage. As one disappointed grower commented, "People who came out of Windsor, what a joke."

Leamington growers also complained that local workers lacked discipline and commitment to the job. One grower commented, "I expect them at eight o'clock in the morning, and they show up twenty-four hours later," and another recollected that "I had two guys who asked me if they can work here. I told them they can start today or tomorrow. Then they asked if they can start tomorrow, and I said, 'No problem.' Tomorrow came and they never showed up for a week." Similar comments were that "Work habits are poor. Just reporting to work is a struggle. To get people to come and if they don't come that they would call and tell you they are not coming. They are not a dependable work force." Another complaint was that "With the local labour what happens is that on Monday morning they are drunk, whether they show up or not." Similarly, "The guys I get are usually single, and they are usually drunk on Friday and they don't come on Saturday morning."

Mexican Mennonite workers, who have a legal residency status in Canada, enjoy a reputation in the Leamington area as hardworking and reliable. Some growers have enjoyed a steady supply of workers recruited through social networks in the Mexican Mennonite community. It is estimated that five hundred Mennonites are employed in the greenhouse industry in Essex County (Hill 1999b). Among some growers the domestic labour force consists almost exclusively of Mexican Mennonites. The retention rate among these workers is rather high, even though as permanent residents in Canada they are free to change jobs. Many prefer working in the field in summer, when it gets too hot in the greenhouses. Some leave for field jobs without notice when it is too late to request offshore labour. As one grower commented, "The Mexican Mennonites, the biggest problem for us is that ... they work for you in the winter, and then in the spring they want to go in the field, so they don't want to work for you full-time. And we can live with that if we know in advance that that's what the rules are. But so often they tell you today that tomorrow they are not going to be there and you can't find people. I can't even apply for offshore to supplement that."

Mexican Mennonites also leave their farm jobs when they find jobs with better pay. One grower reflected: "Do you know that in Windsor there is a lot of shops which have opened up in the last few years, by Mennonites, the ones that come from Mexico. And you go look at their shop. There are seventy-five people working there from Mexico, 70 percent of Mennonites working there. That's the way to go if you get more money, if you get the benefit. Well, we can't give this money in the farms." Another grower observed that "With the expansion of the greenhouse industry there are hundreds of them working for construction companies, house manufacturing. A lot of them started their own welder's shops, little custom shops that cater to our industry. So a lot of them wouldn't come to us."

Since many Mennonite families return to Mexico for a month or two, one grower found himself in a bind. He had two Mennonite families working for him, half his labour force, and suddenly they decided to return to Mexico. He had always relied on local labour, but when they quit he applied for four offshore workers. He described Mexican Mennonites as "somewhat unreliable in the sense that these are a very mobile group of people. On a moment's notice they would pack up and leave. And that's what I am finding now with two of the families. I perceive more problems in the future."

According to one grower, it is hard for farmers to rely on Mennonite women, because they quit paid work when they start having children. In general, as another grower commented, Mennonite women work only when their families are in an economic bind: "Female employees in the case of the Mennonites [work] to make enough money just to get them going again, and then they go back to stay at home with their families. So they may work for 6 months, for 8 months and that may bring enough money for the family so that they can say, 'I am not going to work anymore.' And a few months down the road they may come back and ask for their job back. We have a lot of people who are quitting and rehiring throughout the year. Because it's just this aspect to keep their families above water. Most of the people, they don't want to be working, especially women, some of them have five, six kids."

In sum, despite their reliability, Mexican Mennonites are free to quit their farm jobs when they find better paying ones or when it gets too hot inside the greenhouse or when they have children or when they decide to go back to Mexico. This freedom upsets the greenhouse growers, who maintain that they require labour that is available to work on demand, because the crops are a perishable commodity. Growers may lose much money if crops are not picked on time. The following are some quotations from growers that speak to this issue:

It's like this. When a doctor has to deliver a baby, someone has to be there, a doctor, whoever. And when these vegetables are ready, you've got to pick them. If you don't they'll die. Nine months of work and nothing. Tomatoes and cukes, they don't wait for holidays and Sundays. When the weather like this pushes them, we have to get it out. It's not just a Sunday or a holiday but it's a day the crop has to come out. If you do not have the help you can just lose one picking. And it's not just a thousand dollars, it can go into thousands of dollars. A lot of the farmers, we can't afford that. There is just no way.

And the nature of our business, working with perishable goods, it can't wait. If you don't get the work done, the things will rot. As far as trimming plants and taking care of the plants, pruning them, that needs to be done on time or the crop falls apart.

With tomatoes they ripen seven days a week. They don't just ripen from eight o'clock Monday morning till 4 o'clock on Friday. They ripen all the time.

If it was a piece of wood, piece of steel, yes, I could leave it outside until tomorrow. But with vegetables, if you don't take care of them today, tomorrow it's too late.

I think we need reliability. We can't risk. The crop is so sensitive. The day or two off can do a lot of damage. We can't afford to be without help for a short time.

The reliability part is key. We have to harvest tomatoes. You can't wake up in the morning and find nobody around. Days like today, it's so hot that if you miss some picking you can pretty much throw everything away, because they'll be over-ripe. You have to have reliability. You have to have a core group that it can either rain or shine, whatever.

The problems Leamington greenhouse vegetable growers have had recruiting and retaining reliable workers have made them dependent on offshore labour. Of the 1,026 positions filled in the forty vegetable producing greenhouses surveyed in this study, 418 were taken by offshore workers, 391 of whom were Mexican.

Only three vegetable growers interviewed in the study did not hire offshore labour. They owned small, one- or two-acre farms and found that their labour needs were met sufficiently with local labour. Yet even two of them were planning to request offshore labour in the following year. Only one grower, who employed two local workers, was not interested in the offshore program, because of its high transportation and housing costs.

Greenhouses producing flowers and plants had been excluded from the offshore program until 1999. When the research for this book was conducted, none of the growers operating these greenhouses hired either Mexican or Caribbean workers. Their inclusion in the program is likely to change this situation, although they do not suffer from labour shortages to the same extent as vegetable-growing greenhouses. Since it is not as hot in flower and plant greenhouses and the working conditions are generally better in them, they are more likely to attract local labour. However, certain advantages associated with offshore workers (discussed in chapter 7) would probably encourage these greenhouse operators to hire them.

The Leamington greenhouse industry keeps expanding, and so do its labour demands. Consequently, it is unrealistic to argue, as Nick Ingratta, chair of the Ontario Greenhouse Vegetable Producers' Marketing Board, did in his 1998 annual report, that greenhouse operators need to decrease their reliance on offshore labour and increase their dependence on Canadian labour (OGVPMB 1998, 3). The greenhouse industry has become a year-round business that requires a full-time labour force. Stability of employment could make jobs in the greenhouse industry more attractive to local workers, who would not have to worry about being laid off during the slack season. However, the demand for labour in the winter months is still significantly lower than in the summer. Furthermore, as mentioned, very few people would agree to work in greenhouses in the summer, because of the heat. It will therefore continue to be necessary to recruit offshore labour for these months.

In the last few years technological changes in the greenhouse industry have allowed for some improvements in the working environment. As one industry analyst observes, "The introduction of bumble-bees, nutrient-film fertilizer systems, water fumes in the packing sheds, mobile elevated platforms and Toma hooks have all contributed to a more worker-friendly workplace. Also, the adoption of biological integrated pest control systems has reduced the use of pesticides and minimized the exposure of workers to toxic chemicals" (Stevens 1996, 16). However, the improvements in the working conditions in the greenhouses have not been sufficient to attract local labour. And that is why many Leamington growers became alarmed by Premier Harris' proposal concerning workfare for farm labour, fearing the negative impact of this proposal on the offshore program. As Charlie McLean, president of the Essex County Federation of Agriculture, put it, "If the offshore program is reduced or cut out because of this and it doesn't pan out ... we can't wait three or four weeks to get these offshore workers back" (Schmidt 1999).

THIS CHAPTER DESCRIBES THE MULTIPLE
HIRING PROBLEMS FACED BY LEAMINGTON
GROWERS TO HIRE LOCAL WORKERS.
EITHER THEY ARE TOO UNRELIABLE,
DO NOT ACCEPT THE WORKING CONDITIONS,
OR ARE NOT WILLING TO PUT IN
ALL THE HOURS REQUIRED BY GROWERS
FOR AN ADEQUATE HARVEST WHEN
FRUITS OR VEGETABLES HAVE RIPPENED.
BECAUSE OF THIS, GROWERS HAVE COME TO
DEPEND ON OFFSHORE WORKERS. IN
ORDER TO HAVE ACCESS TO THESE, THOUGH,
GROWERS MUST DEMONSTRATE INABILITY
TO HIRE LOCALS. ~~TO GET PERMIT TO~~

4 Vulnerable Labour in a Vulnerable Sector

As discussed in chapter 2, demands for agricultural wage labour have been rising since the 1940s. At the same time, Canadian growers have experienced significant problems trying to recruit farm help among domestic workers. But the biggest problem that has plagued Canadian food growers has been a failure to retain hired help (see chapter 3). Many factors contribute to the high turnover rate among farm workers. Among them are relationships with the employer, the method of recruitment, the presence of a signed contract, and the origin of the workers. But the most significant factors affecting turnover are working conditions and wages. Farm workers receive very low wages; their jobs are dull, difficult, and dangerous; they are excluded from provincial labour legislation; and they are not allowed to form unions. Very few Canadians are willing to perform farm labour under these conditions at existing levels of remuneration. While government officials have blamed the growers for their failure to retain farm workers, the growers, on the other hand, have attributed their problems to external circumstances, such as the availability of higher paying jobs in urban industries. Researchers have related the failure of the growers to provide decent wages and acceptable working conditions to the vulnerability of family farming: farmers find themselves in a "cost-price squeeze" between the low prices they have to charge for their produce and the high cost of various farm inputs.

THE PROBLEM OF HIGH TURNOVER

Shortages of reliable labour have plagued Canadian farmers for several decades, and the high turnover of farm workers has posed a major challenge to many food growers. Several researchers have tried to identify the factors that contribute to labour stability. First, as pointed out by numerous studies cited by Smit, Johnston, and Morse (1984, 17), the relationships between employers and their workers and the management practices of the employers affect the turnover rate significantly. When farms are smaller or the relationship with the employers is more personal, the turnover rate tends to be smaller. Second, recruitment strategies also play a significant role: workers recruited through personal contact tend to stay on the job longer (37). Third, a signed contract eliminates confusion regarding terms and conditions of employment and is an essential part of any attempt to reduce employee dissatisfaction and turnover (41). Fourth, the workers' place of origin is related to their propensity to stay on the job. In their study of turnover among workers employed in the tobacco industry, Smit, Johnston, and Moore observed that Ontario workers who were not local were much more likely to terminate their employment early than were both local and foreign worker groups (61). Local workers are more familiar with the rigours of tobacco employment and therefore are more likely to persevere than nonlocal Ontario workers. Foreign workers are less mobile in comparison to nonlocal Ontario workers, because they are assigned to a particular employer by a contract.

Notwithstanding the factors listed above, low agricultural wages and dismal working and living conditions contribute the most to the high turnover among farm workers (Ward 1974, 19). In 1974, for instance, wages in manufacturing exceeded agricultural wages by 1.88 times and in construction by 2.77 times (Dawson and Freshwater 1975, 10). In virtually all provinces, agricultural workers have been exempt from provisions concerning holidays with pay, vacations with pay, minimum wages, hours of work, overtime pay, and pregnancy leave coverage (Dawson and Freshwater 1975, 27; Wall 1994, 66; Parr 1985, 103; Mitchell 1975, 28; Stultz 1987; Shields 1992, 250–1). It is not unusual for farm workers to be paid less than a minimum wage (Mitchell 1975: 28; Dawson and Freshwater 1975, iii).

The Ontario Labour Standards Act distinguishes between farm workers and harvesters, but this makes little difference in practice.

A farm worker is employed to perform such tasks as planting crops, cultivating, pruning, feeding and caring for livestock, and transporting produce to market. A harvest worker is employed to bring in crops of grain, seed, fruit, vegetables, or tobacco for marketing or storage. Farm workers are covered by only some of the basic minimum standards of the act, including pregnancy and parental leave provisions, notice of termination, severance pay (if they qualify), equal pay for equal work, and regular payment of wages. (The latter provision means that wages are paid out on a regular day at a regular place and that a written wage statement showing details of the hours worked, the wage rate, and any deductions is provided for each payment.) Harvesters, on the other hand, are also covered by the provisions concerning minimum wages, public holidays, and vacation pay, *but* only if they have been employed as harvest workers (and not as farm workers) for at least thirteen weeks. This specification makes it difficult for harvest workers to claim these benefits, since harvest seasons generally do not last that long. Neither harvest nor farm workers are covered under the overtime pay provisions.

Work-related accidents are frequent on farms, and together with extensive use of toxic chemicals and exposure to various airborne dusts and animal-borne diseases, they contribute to precarious working conditions that have been classified as the most dangerous after conditions in mining and construction. Pesticides used in agriculture are not properly tested, and in some cases they are used in spite of their known detrimental health effects. Additionally, farm workers are exposed to commercial solvents and strong sunlight and heat. Finally, they also suffer from mental illnesses because of insecure and depressing working conditions; racial subordination; material, social, and environmental deprivations; and uprootedness and consequent disruption of family and community ties (Ward 1974, 9; Bolaria, Dickinson, and Wotherspoon 1991, 399–401, 403, 404). Because they are covered by the provincial acts, agricultural workers have no legal recourse in matters of health and safety. Even though the British Columbia Social Credit government did decide to extend Workers' Compensation Board health and safety regulations to farm workers in 1982, it reversed its plan in 1983 (Shields 1988; Bolaria 1992, 242).

In every Canadian province except British Columbia it is illegal to form agricultural workers' unions (Mitchell 1975, 29; Stultz 1987). Bill 91, the Agriculture Labour Relations Act, introduced by the New Democratic Party government in Ontario in the early 1990s, which would have allowed unionization of the family farm, was repealed in 1995 by the Conservative Party. There have been some attempts to organize farm workers in British Columbia and in other provinces, but they have been largely unsuccessful. The Canadian Farmworkers' Union was recognized in British Columbia in 1980, almost a decade after the idea of forming a farm workers' union was conceived. Among the CFU's objectives were complete abolition of the abusive farm labour contracting system, legal changes governing the handling of pesticides, and the inclusion of farm labour in the province's workers' compensation system and minimum wage laws, as well as easy access to drinking water and toilets and to daycare (Tatroff 1994, 25). Despite some success, the CFU failed to significantly improve the working conditions of rural workers. In 1994 Tatroff described the conditions under which farm workers worked and lived as follows: "Farmworkers are still being abused, they're still being housed in filthy shacks; they're still being packed like sardines into beat-up old school buses and driven to fields where they're forced to breathe in toxic pesticides; they're still not receiving the minimum wage; they're still being short-changed at the weight-scales; and families ... are still being cheated out of their meagre earnings by greedy farm labour contractors" (27). He observed that there was no running water on the farms, no daycare for children, and when housing was provided, it was "overcrowded, filthy and expensive" (23). Other researchers (cited in Bolaria, Dickinson, and Wotherspoon 1991, 403–4, and Bolaria 1988, 117–19) portray a similar picture.

Even though the CFU also spread to Ontario, its presence there was short-lived. In 1981 it launched an organizational campaign among tobacco workers in southwestern Ontario. Tobacco farms were chosen because they are among the most prosperous in the country, and work in the tobacco sector is most akin to industrial work: it requires skilled labour, and the harvest period is longer than with other crops (Shields 1992, 252). Yet, the organizing efforts did not come to fruition, and by May 1983 the CFU had closed down its office in Ontario (Stultz 1987, 295). It has been difficult to organize farm labour for several reasons. First, farm workers live predominantly in rural areas and have very little contact with other farm workers outside their place of employment (Stultz 1987, 294). Second, solidarity among farm workers is inhibited by the paternalistic relations the workers develop with their employers (Stultz 1987, 295; Shields 1992, 250; Wall 1992, 268). As Wall contends, "such dependencies help to cement farm workers into personal relations of unequal exchange" and "undermine the development of solidarity among disadvantaged workers." Even if Ontario farm workers were legally allowed to form work associations, "allegiance to their employers might become a major obstacle to the success of such groups" (1992, 268). Third, lack of transportation and the requirement of working long hours prevents workers from participating in activities unrelated to their work. Fourth, farm workers often live on the employer's property and are virtually inaccessible to outsiders (Stultz 1987, 294–5). Fifth, many farm workers are recent immigrants who speak different languages, and their linguistic, as well as cultural, differences hamper communication between them (Stultz 1987, 295; Shields 1992, 250). And finally, it is hard to interest workers in joining a union when many of them stay on the farm for only one or two seasons (Tatroff 1994, 24).

The CFU has achieved the most in British Columbia, the province that relies more heavily on paid labour in agriculture than any other Canadian province: about half the farms use at least some wage labour (Shields 1992, 249). Another contributing factor is the relative ethnic homogeneity of British Columbia farm workers in the Fraser Valley, the majority of whom are from the Punjab region of India (Stultz 1987; Shields 1992, 248–51). But even there the SFU seems to have been losing ground. Whereas in the mid-1980s there were 1,400 SFU members, in 1994 there were as few as 407 (Tatroff 1994, 26).

WHO IS TO BLAME?

Government officials have blamed the growers themselves for the dismal conditions they offer their workers. The 1973 report by the task force of the federal Department of Manpower and Immigration provides an illustration:

> We visited one cucumber and tomato grower north of Chatham who has a very large and modern operation. After an angry tirade about Canadians not wanting to work, and other familiar themes, he took us to his cucumber fields, where workers were then picking small pickling cucumbers to fulfill a contract. This grower operates with his pickers on a 50–50 split

of the selling price basis, under which, naturally, the worker's efforts are reflected in his earnings. At that time, the selling price was $2.30 per basket, which means that the worker would earn $1.15 per basket. However, this grower neither pays any wages nor advances any money until the picker has been with him for a week, after which time the worker is paid 40 cents per basket, the remainder being "banked" by the grower until the season is over and the picker leaves. Most transient pickers arrive with very few assets, if any, and this arrangement for pay means that for one week, plus whatever period of time before that, the picker may not have eaten. This worker has absolutely no money even for basic food supplies. It is not hard to understand why, under these circumstances, there is a high turnover.

We visited a cherry and apple grower, an extremely intelligent and thoughtful man. He has no manpower problems out of the ordinary, probably because he treats his employees as human beings rather than as chattels, as do some other growers. When the subject of Mexican workers arose, he informed us very plainly and firmly that he strongly disapproves of people in these unfortunate circumstances being brought into Canada "like rats (his word), and treated and paid like rats." He stated that he does not object to the unit costs of production, including manpower costs, as long as he is not the only one paying a higher, or, in wage terms, a decent rate. What he does object to strenuously is taking his produce to market and having a competitor coming in at the same time and undercutting his prices because the competitor has employed exploited labour in the process. (Department of Manpower and Immigration 1973, 25–6)

In 1995 and 1996 a series of meetings was held between representatives of Human Resources Development Canada (HRDC), Human Resource Centres Canada (HRCC), and growers at which several HRCC staff expressed their belief that the less-than-reliable workers and the high turnover rate are growers' problems, insofar as the growers fail to provide adequate wages and working conditions (FARMS 1996, 3–4).

Growers, on the other hand, blame the situation on external circumstances. As one farmer interviewed by Larkin and Manning put it, "How many people wanna do stoop labour ... like bending over and picking tomatoes, picking cucumbers ... Not everybody. They'd say: 'Hell ... I'd rather go on unemployment or you name it, but I'm not gonna pick it!'" (Larkin 1990, 57). Leamington greenhouse growers blame their labour retention problems on hot temperatures in the greenhouse: "When it gets hot, when it's ninety outside local people leave. So what are you gonna do? When it's snow outside you can attract people to work inside 'cause it's warm. It's nice to work like this when it snows. When it's ninety, what are you gonna do?"

Leamington growers also realize that higher wages elsewhere make farm jobs unattractive. One grower observed that "The biggest competition we've had, especially for male workers, are factories in Windsor. A lot of our workers are immigrants from different countries. When they first come into the area, farm labourers are the easiest to get hired quickly and start making money. And once they've learned the system and how it works and that there are other jobs elsewhere which start at eight, nine dollars an hour, it's been pulling a lot of people out of the area, especially male workers." Another grower commented on how higher wages in the automotive sector and low unemployment rates are responsible for their problems, "You have to realize that we are in Essex County, close to Windsor, with big jobs in the automotive sector, with all the spin-off jobs in the automotive sector, they are paying from twelve to twenty plus dollars per hour. How do you compete against that – with all the benefits these jobs offer? This is our biggest problem. Because we are in the area where unemployment is almost nothing right now, 5.7 percent unemployment, it's almost no unemployment. Whereas the greenhouse operations out East, they don't have a problem with local labour. Actually, they are so scattered, you may have an acre here, two acres there. They are just all over the place, and they just get local help within their community."

THE "VULNERABILITY" THESIS

In the academic literature farm workers are considered to be the "most exploited segment of the working class in Canada," because of the vulnerability of agricultural enterprises (Bolaria, Dickinson, and Wotherspoon 1991, 403). As mentioned earlier, several authors maintain that Canadian family farming is vulnerable to a cost-price squeeze (Anderson and Daniel 1977; Mitchell 1975; Winson 1992; Satzewich 1991; Shields 1988, 1992). While Canadian farmers pay increasingly high prices for farm inputs, such as fertilizer and seed, produce prices are kept low. The economic expansion that Canada enjoyed in the 1960s intensified this cost-price squeeze. Between

1961 and 1970 farm wages increased by 91.3 percent; the cost of inputs by 22.9 percent; the cost of interest and taxes by 51.8 percent; and the cost of living in rural areas by 25.7 percent. At the same time, agricultural prices increased by only 26.7 percent between 1961 and 1968 (Satzewich 1991, 153).

In the 1970s land prices were so high that they constituted "the greatest single drain on capital." With the advent of capital-intensive agriculture, competition for scarce land intensified and land prices skyrocketed. As a result, land represented the lion's share of capital investment in farms at that time (Mitchell 1975, 21, 22). Ghorayshi (1986, 147) points out that the average amount of capital invested in farms kept increasing, placing a significant burden on farmers who had to obtain the required capital and who consequently became extremely dependent on many public and private credit institutions. By the mid-1980s many farmers had accumulated high debts that threatened their very existence, especially in the context of declining prices and land values (Bolaria, Dickinson, and Wotherspoon 1991, 402). The report of the House of Commons Standing Committee on Agriculture of 1988 (cited in Bolaria, Dickinson, and Wotherspoon 1991, 402) estimated that as many as one-third of farmers were "experiencing rising levels of financial stress in spite of massive government assistance." It further suggested that for farmers who entered farming during the 1970s or later the debt was particularly high, making it difficult for them to stay in business. As a result many farmers were foreclosed by lending institutions. In addition to the costs of land, elevated machinery prices made it difficult for many farmers to stay afloat (Mitchell 1975, 18–24). For those farms that did manage to stay in business, the cheap food policy pursued by the Canadian state made it very difficult to offset rising production costs. The government set low tariffs on imported low-priced fruit and vegetables from such areas as South Africa, California, Mexico, South Korea, and Australia, creating unfair competition for Canadian farmers (Satzewich 1991, 67; Anderson and Daniel 1977, 13–14; Warnock 1978, 113–17; Shields 1992, 258–61). By comparison with other countries, such as the United States, Mexico, and Australia, "Canadian growers have to deal with a colder climate, the ability to produce only one crop a year, a shorter growing season, and much higher land costs" (Warnock 1978, 16).

The concentration of ownership in the implement producing, food processing, and retail food industries has intensified the cost-price squeeze. Large implement-producing companies set elevated monopoly prices on farm inputs. Consolidated food processing companies, many of which are owned by such U.S.-based transnational companies as Heinz, Green Giant, Campbell's Soup, Del Monte, and Libby's (Warnock 1978, 106–7), control the food market and set low prices (Satzewich 1991, 65), despite attempts by product marketing boards to negotiate fair prices (Winson 1992, 144–5; Warnock 1978, 111–12) . Large processing plants further undermine Canadian farming by importing low-priced food. Similarly, chain store companies are not dependent on domestic suppliers, since they have access to low-priced imports (Anderson and Daniel 1977, 16). The intensification of foreign penetration in the food processing industry since the signing of the Free Trade Agreement with the United States in the late 1980s has posed significant problems for Canadian farmers (Winson 1992, 129).

The concentration of ownership in the food processing industry has also contributed to an increasing social differentiation among Canadian farmers. Growers in Canada are integrated into the food industry though contractual linkages. There is evidence to suggest that larger processors generally award contracts to the larger, more highly capitalized growers, in order to secure a permanent supply of food and save on administrative costs. The disappearance of small processors has therefore led to the demise of small growers as well (Winson 1992, 153–4).

Concerns expressed by various academics about the fate of Canadian family farming are mirrored in the statements made by farmers themselves. Writing on behalf of the Essex County Associated Growers, Bezaire describes the plight of Ontario farmers: "Growers of early fruits and vegetables must sell on uncertain markets. There is no way of limiting production to a basic level achieved by production-planning in other businesses. Growers produce a perishable product which must be sold on an open market. A five percent of excess production can often mean prices which are below the break-even point. In many instances, the over-production which plagues the producer may have been produced outside the country, but finds its way to the Canadian fresh fruit and vegetable market to add to the element of risk in an already risky venture" (1965, 16). One of the farmers interviewed by Larkin and Manning explains the consequences of the cheap food policy for Canadian farmers in the following manner: "In the food chain, a lot of consumers don't realize that back in ... the '60s the consumers were paying .48 out of a dollar for their living, and today you're only paying .15 out of a dollar. And by the same token, a lot of the producers that are producing corn, it's the same price it was 50 years ago ... How would you like to work for wages that were paid 50 years ago? And you say you've gotta compete ... And all the blame's put on the farmer" (Larkin 1990, 57–8).

Leamington greenhouse vegetable growers have also complained about the vulnerability of farming. As one grower commented at the end of an interview, "What you didn't ask is why we don't pay more? There is not enough gravy to go around. The industry would cease to function if farmers had to pay higher wages." Another said, "Now you gonna say, maybe, you've got to pay them more. But remember, it's a farm. I don't know how much higher we can go. So ..." Similarly, "We can't pay them much because we never know what we're going to earn. There is no price set per cucumber or per pound of tomatoes. It fluctuates. So we have a set wage."

In sum, Ontario farmers must deal with the uncertainties of the market and climatic conditions, harsh international competition, and rising input prices, all factors that are outside their control. But there is one cost that they *can* control, namely, the wages paid to workers. The incentive to keep wages down is related to increases in other capital investments. Since labour costs make up almost 15 percent of all operating expenses in Ontario agriculture (and an even higher share in the labour-intensive sectors, such as fruit, vegetable, and tobacco production), it is in the farmers' interest to prevent their increase. Shields explains why legislative control over agricultural work would not only add significant costs to the farmers but would also endanger the very survival of agricultural production. With respect to health and safety coverage, he observes, "Workers would have to leave the fields during pesticide spraying; it could be days after spraying before workers could 'safely' re-enter the fields and continue their labour ... In a sector as sensitive as agriculture, where even short-term delays can often mean the difference between success and disaster, such considerations become extremely important" (Shields 1992, 255, 256). Given the perishability of the crops, any restriction on hours of work at harvest time could pose a threat to the economic viability of a farm. Furthermore, if agricultural labour was permitted to organize and stage strikes, entire harvests could be lost. Finally, increased labour costs would place Canadian farmers at a disadvantage relative to their southern neighbours (263–5).

SUMMARY

Very few Canadian workers are willing to accept farm jobs, and those who do take them do not stay. There are several reasons why the turnover rate is so high, but the most significant are the dismal working conditions and the low wages. Farm workers are exposed to numerous health hazards, including pesticide poisoning, farm accidents, dust-borne diseases, and emotional stress, and they are not covered by any labour acts. At the same time, the wages paid to them are too low to make these conditions tolerable. There is therefore no incentive for them to stay on the job. Although harvesters in Ontario, as opposed to farm labourers, are entitled to some protections provided in the Labour Act, they are eligible only after they have been employed as harvesters for thirteen weeks. This requirement makes it very hard for many harvesters to claim their benefits. In no province but British Columbia are farm workers allowed to organize to demand improvements in their working conditions.

It has been argued that if farmers were to improve the wages and working conditions of their workers, they would hardly be able to survive, since they have been rendered vulnerable by the government's cheap food policy and the increasing prices of inputs. This picture is too general, however. As Winson (1996, 92–3) has pointed out, farmers have undergone a process of social differentiation, and a privileged stratum of farmers who retain a disproportionate control over land and rural resources has emerged: not all farm sectors are equally vulnerable. The Leamington greenhouse industry is one sector that has experienced significant growth. It is the subject of the next chapter.

5 The Greenhouse Industry in Leamington

with CAROLYN LEWANDOWSKI

As mentioned at the end of the previous chapter, it has been argued that many Canadian growers of fruits and vegetables cannot afford to increase wages and provide better working conditions because of their vulnerability to increasing input costs and the low prices they are forced to charge for their produce. However, not all Canadian growers are vulnerable. Some have enjoyed significant profits over the years and have been able to expand their production. Many Leamington greenhouse vegetable growers are among them. In the last decade both production levels and sales have increased in the greenhouse vegetable sector, despite decreases in prices. The healthy profits made or expected by many Leamington growers have contributed to the extraordinary growth in greenhouse construction, which can be attributed to several factors, including an improved product and cost-reducing technologies. However, as mentioned, this growth has been accompanied by social differentiation among the greenhouse growers. While some have come to occupy the rank of the wealthiest growers in the country, others are still owners of small, family-run businesses struggling to stay afloat.

THE IMPORTANCE OF LEAMINGTON

Canada is a major player in greenhouse vegetable production. In 1998 the industry earned $209 million and covered 767 acres (Khosla

Table 5.1
Canadian Greenhouse Vegetable Production, Area in Acres, 1 March 1999

Province	*Tomatoes*	*Cucumbers*	*Peppers*	*Lettuce*	*Total*
Ontario	539	271	76	7	893
Quebec	165	42.5	5	37.5	250
British Columbia	185	40.5	136.2	6.3	367.5
Alberta	12.5	46	6	1	65.5
Total	901.5	399.5	223.2	51.8	1,576

Source: Whitfield and Papadopoulos (1999, 6).

Table 5.2
Greenhouse Vegetable Sales Estimates, 1997, in Millions of Dollars

Province	*Tomato*	*Cucumber*	*Pepper*	*Lettuce*	*Total*	*Percentage*
Ontario	71.1	35.9	11.5	2.7	121.2	47.1
Quebec	27.8	3.2	0.0	3.8	34.8	13.5
British Columbia	33.6	11.8	30.8	5.0	81.2	31.6
Alberta	3.4	10.0	1.2	0.0	14.6	5.7
Nova Scotia	2.4	1.7	0.0	0.0	4.1	1.6
New Brunswick	1.1	0.0	0.3	0.0	1.4	0.1
Canada	139.4	62.6	43.8	11.5	257.3	100.0

Source: Whitfield and Papadopoulos (1999, 6), based on Statistics Canada, Catalogue no. 22-202-XPB, 1997.

1998). With greenhouse construction continuing in 1999, it was expected to expand to 907 acres. The main greenhouse vegetable crops in Canada are tomatoes, cucumbers, peppers, and lettuce. More than half the area of Canadian greenhouse tomato production and almost two-thirds of the area of greenhouse cucumber production are located in Ontario (table 5.1). Ontario is also a leader in sales, especially of cucumbers and tomatoes (table 5.2), and it is a net exporter of greenhouse tomatoes and cucumbers to the United States. According to the Ontario Greenhouse Vegetable Producers Marketing Board (OGVPMB), 70 per cent of the tomato crop for the spring of 1998 was exported to the United States, including to such regions as California and Florida. At the same time, the Ontario greenhouse industry is the main supplier of fresh vegetables to markets in Eastern Canada (Whitfield and Papadopoulos 1999).

In Ontario, the major producing area is southern Essex County, the area in and around the town of Leamington. In 1999 the Leamington greenhouse industry, with 719 acres, had the largest

Table 5.3
Acreage in Greenhouse Vegetable Production

Crop	*Leamington 1988*	*Leamington 1999*	*Rest of Ontario, 1988*	*Rest of Ontario, 1999*
Tomatoes	140	445	51.7	94
Cucumbers	109	215	37.5	56
Peppers	0	47	0	29
Lettuce	0	0	0	7
Total	249	707	89.2	186

Source: Whitfield and Papadopoulos (1999, 8–9).

concentration of greenhouse vegetable production in North America (Whitfield and Papadopoulos 1999). Located within fifty kilometres of the U.S. border, Leamington has the advantage of allowing its growers relatively easy access to markets in the United States. The Leamington district contains 72 percent of South Essex greenhouses, with 24 percent of the greenhouses being located in the Kingsville area and the remaining 4 percent in the Cottom, Ruthven, and Wheatley areas (Greenhouse Sector 1996). In 1999, about 80 percent of the area in tomato and cucumber production and more than half the area in sweet pepper production was located in Leamington (Khosla 1998). Further, the total area in greenhouse vegetable production in 1999 was 2.9 times the area in production in Leamington, whereas in the rest of the province it was only 2.1 times the area in production in 1988 (see table 5.3). In 1999, 719 acres were in greenhouse vegetable production in Leamington, including the area of new greenhouse construction (Khosla 1998).

In 1999 and 2000 the towns of Leamington and Kingsville became concerned with the impact of greenhouse expansion on water pressure for residents and issued a moratorium on greenhouse development until a water impact study could be completed (Hill 1999a, 2000b). However, since it has been estimated that an acre of greenhouses generates about $325 annually in municipal taxes in Kingsville, compared to $7.65 for an acre of vacant farm land, instead of attempting to curb greenhouse expansion, both town councils proposed that greenhouse growers contribute to the cost of installing more watermains, so that the industry could continue to expand (Hill 2000a, b).

Table 5.4
Greenhouse Vegetable Operations Larger than Ten Acres, 1 March 1999

Name	*Location*	*Acreage*
Mastron Enterprises Ltd	Leamington, ON	57
Houweling Nurseries, Ltd	Delta, BC	48
Veg Gro Inc.	Leamington, ON	36
DiCiocco's	Leamington, ON	32
Canagro	Delta, BC	31
Amco Farms	Leamington, ON	30
Mucci Bros.	Leamington, ON	30
Les Serres du St Laurent	Portneuf, QC	30
Cervini's	Leamington, ON	25
Delta Pacific	Delta, BC	25
St Davids Hydroponics	St Davids, ON	24
Gipaanda	Delta, BC	24
Hazelmere Greenhouses Ltd	Surrey, BC	21
Suntastic Hothouse	Exeter, ON	20
Howard Huy Greenhouse	Leamington, ON	20
Double Diamond Acres Ltd	Leamington, ON	14
Hydro-Serre Mirabel	Mirabel, QC	12
Hillcrest Farms	Leamington, ON	10
Total		489

Source: Whitfield and Papadopoulos (1999, 7).

The significance of Leamington greenhouse vegetable producers is also demonstrated by the list of the top Canadian producers in table 5.4. Of the largest 18 producers (those with over 10 acres in production), 9 are located in the Leamington area, constituting 52 percent of "top-producer" acreage. Furthermore, the producers listed in table 5.4, who constitute just 0.5 percent of all Canadian greenhouse producers, account for 31 percent of the total Canadian acreage devoted to greenhouse vegetable production, indicating a concentration of the industry in a minority of operations in 1999. The nine Leamington producers alone control 16 percent of the total Canadian acreage producing greenhouse vegetables. The significance of this control and the extraordinary importance of Leamington producers is indicated by the revenue that is generated by the greenhouse vegetable industry (table 5.5).

Leamington vegetable greenhouses are larger than greenhouses elsewhere in the province. The average size of a Leamington-area greenhouse in 1996 was 2.5 acres, almost seven times larger than an average Canadian greenhouse (table 5.6). Among the forty

Table 5.5
Value of Ontario Greenhouse Vegetable Production 1998

District	*Value (Millions of Dollars)*
Leamington	180
Rest of Ontario	52
Total, Ontario	232

Source: Whitfield and Papadopoulos (1999, 9).

Table 5.6
Area of Vegetable Greenhouses, by Region, 1996

Region	*Total Acreage*	*Number of Farms*	*Average Size (Acres)*
Canada	1,039	2,903	0.36
Ontario	509	785	0.65
Essex	311	140	2.2
Leamington	237	95	2.5

greenhouse vegetable growers interviewed in this study in 1999, the average size was even higher – 7.7 acres. While some of the difference may be attributed to the overrepresentation of large greenhouses in the sample, it may also reflect the expansion that many farms underwent between the 1996 census year and 1999. According to OGVPMB general manager Denton Hoffman, there has been a significant shift to larger greenhouse operations. Most operations with less than half an acre have disappeared (OGVPMB 1998, 8).

PRODUCTION TRENDS

The greenhouse vegetable industry is a vital sector of Canadian agriculture that has increased its production from an estimated value of $80 million in 1988 to $250 million in 1997 (Whitfield and Papadopoulos 1999). According to the federal minister of agriculture, Lyle Vanclief, the "industry has blossomed into a $90 billion industry in Canada, with exports to other parts of the world reaching $2 billion" (Santos 1998). Nick Ingratta, the chair of the Ontario Greenhouse Vegetable Producers Marketing Board and an Essex County producer, claims that "the greenhouse industry is expanding at a rate of 100 acres per year" (Lackrey 1998). The greenhouse industry in Canada has grown about 20 percent in the

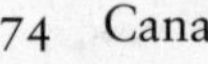

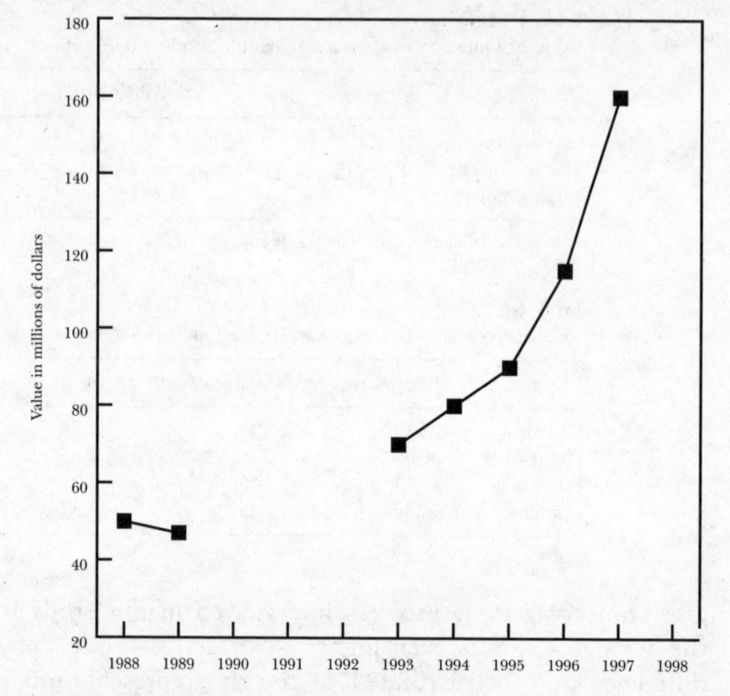

Fig. 1 Ontario greenhouse vegetable production, 1988–98
Note: The lacuna in production values is attributed to a period of "internal destructive competition" that caused a reduction in values and that was alleviated when the OGVPMB regained the right to set prices (res.agr.ca/harrow/hrcghar.htm). No data was available for this period.

past three years, and half that growth has been attributed to Essex County producers (*North Essex News*, 8 July 1998).

The greenhouse industry has also experienced significant growth in the last decade (figure 1). In 1985, yields began to increase rapidly, due to the implementation of new technology and increased automation at Ontario greenhouse vegetable operations. The greenhouse vegetable industry in Ontario also demonstrated strong growth in farm value during most of the 1980s, which was attributable to a corresponding increase in production. Farm value grew steadily from approximately $20 million in 1980 to over $50 million in 1988. Then, because of competition among farmers who tried to undercut each others' prices, it decreased. But since 1991, when the Greenhouse Vegetable Producers' Marketing Board reestablished its control over the price of greenhouse vegetables in Ontario, farm value has been climbing. In 1993 the Ontario greenhouse vegetable industry experienced a good year overall. Greenhouse acreage expanded by

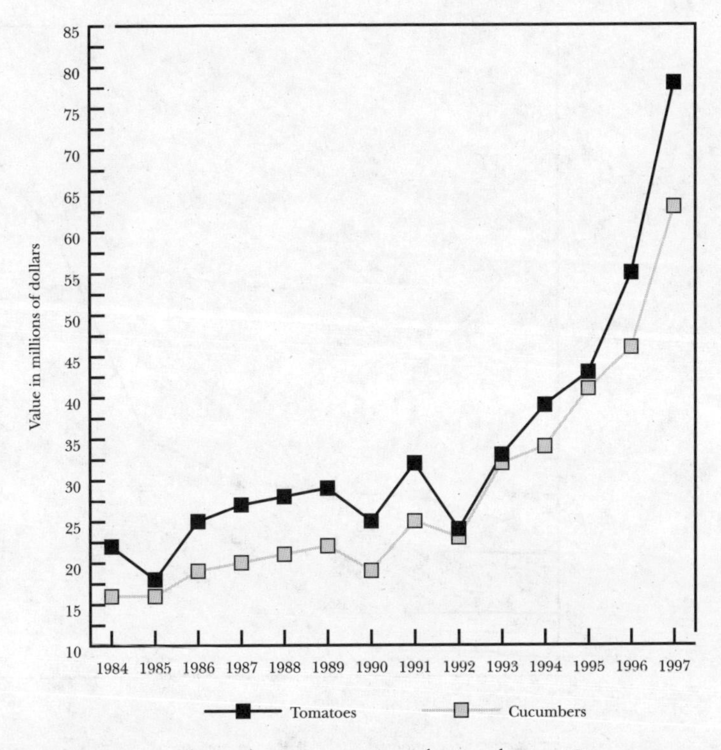

Fig. 2 Ontario production of greenhouse tomatoes and cucumbers, 1984–97
Source: Annual report to the OHCRSC (1997) res.agr.ca/harrow/hrcghar.htm.

approximately 10 percent, mostly in tomatoes. According to industry analysts the value of the greenhouse vegetables produced in Ontario in 1993 reached a record $70 million.* Since then it has continued to grow (Whitfield and Papadopoulos 1999). As can be seen from figure 2, the production of both tomatoes and cucumbers increased almost four times over the thirteen years from 1984 to 1997. Even though greenhouse tomato prices fell in 1992 and again between 1995 and 1997, because of rapidly rising production levels, sales experienced a steady increase between 1991 and 1997 (figure 3).

* There is often a discrepancy between estimates made by industry analysts and those reported by Statistics Canada. According to Statistics Canada, the value of greenhouse vegetables produced in Ontario in 1993 was only $54 million.

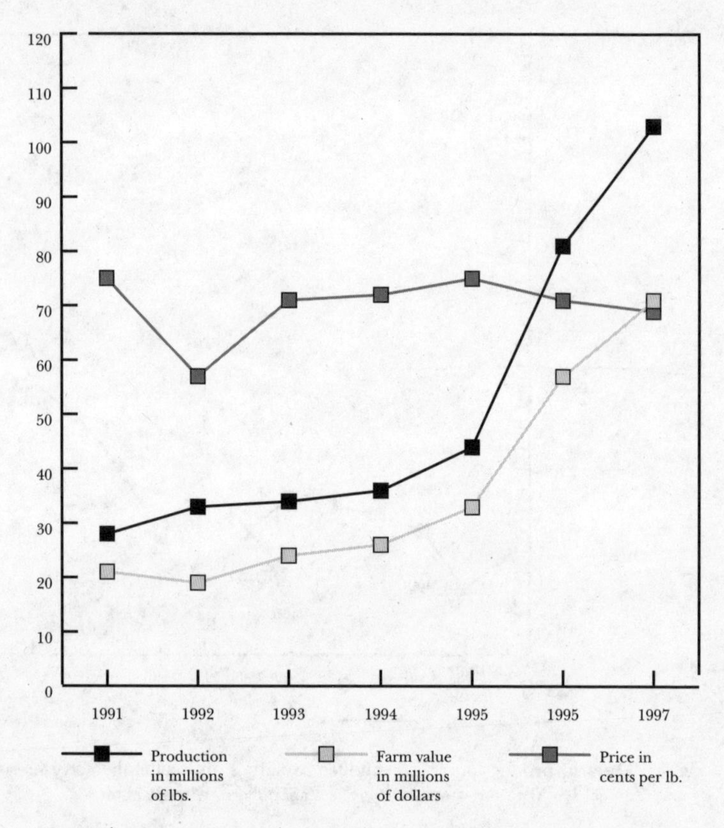

Fig. 3 Ontario greenhouse tomato production, 1991–97

EXPANDED MARKETS

OGVPMB chair Nick Ingratta attributes some of the success in the greenhouse industry to the increased sales and promotion in the United States. As he points out, "co-operative marketing efforts with our sales people have been effective [in introducing] greenhouse tomatoes and cucumbers to our southern consumers" (OGVPMB 1998, 3). Industry analysts are optimistic about the potential for further expansion, because of changes in consumer preferences towards fresh, natural, healthy foods. It is expected that the greenhouse vegetable industry will increase exports to the United States, particularly in the untapped consumer markets of big cities like New York, Boston, Detroit, and Chicago (Whitfield and Papadopoulos 1999).

Even though the Leamington greenhouse industry faces competition from U.S. and Mexican growers, it has nevertheless been able to market its produce quite successfully. For 1999, the Ontario Ministry of Agriculture, Food, and Rural Affairs (OMAFRA) estimated that the United States had 725 acres in greenhouse production, and Mexico 850 acres (Khosla 1999). In some respects Leamington producers have a competitive edge over the U.S. and Mexican greenhouse producers. According to Denton Hoffman, Leamington enjoys slightly reduced costs of production due to the concentration of its operations: "Suppliers are able to provide better prices due to volume and delivery efficiency. The infrastructure for suppliers and marketing is well established and continues to improve over time due to competition. The highest percentage of the industry is located in close proximity to highways that have easy accessibility to the main market centres" (OGVPMB 1998, 8). Denton points out that with respect to two other critical aspects of greenhouse production, namely water and fuel, Leamington is also well positioned. The industry is located around the Great Lakes and thus has a good source of high quality water. And most of the greenhouse operations are serviced by natural gas, a clean fuel that requires less maintenance than previous greenhouse fuels (OGVPMB 1998, 8).

By comparison, production costs are generally higher in the United States. Even though production costs are low in Mexican greenhouses, transportation costs would significantly increase the price to consumers (Sullivan and Garleb 1996, 20).

Until now the Leamington greenhouse industry has not been threatened from other sources. Greenhouse vegetables grown in Spain, Israel, Italy, Greece, Turkey, Morocco, and Egypt are sold predominantly in Europe and do not pose serious competitive threats to Leamington growers (Sullivan and Garleb 1996, 24). However, some Leamington growers interviewed for this study did express concern about greenhouse tomatoes imported from the Netherlands. Despite the high transportation costs, Dutch greenhouse tomatoes could be marketed in Canada cheaply because of state subsidies extended to Dutch greenhouse growers. Thus, although the Leamington industry has so far not been undermined by the importation of tomatoes from the Netherlands, this situation may change in the future.

TECHNOLOGICAL CHANGES

On 29 June 1998 the Greenhouse and Processing Crops Research Centre (GPCRC) at Harrow, Ontario, held a public relations gathering to celebrate the opening of its \$2 million greenhouse expansion, which was made possible by a joint venture between the Ontario Ministry of Agriculture and Public Works and Government Services Canada. On hand was Nick Ingratta, OGVPMB chair and local producer, who stated that greenhouse cucumber and tomato growers have "donated close to \$1 million to the Harrow centre over the last 15 years ... because the work being done at Harrow is critical to the greenhouse industry" (Lackrey 1998). The work accomplished by the Harrow centre is in fact so critical to the success of the industry that the growers' donation must be considered a resource well invested. A research priority of the Research and Services Committee, a subcommittee of the GPCRC staffed by professional personnel from the centre and area greenhouse vegetable producers, is to "develop and evaluate production systems that improve profitability of greenhouse vegetable crops" (Whitfield and Papadopoulos 1999). It seems that the centre has been successful in achieving this goal.

The greenhouse vegetable research team, which studies plant physiology, plant pathology, entomology, and greenhouse environment control, is the largest of its kind in North America (Whitfield and Papadopoulos 1999). Research originating at the Harrow station has resulted in improved technologies that have been disseminated to vegetable producers. A shift from glasshouses to double-polyethylene houses has resulted in substantial savings in heating costs (up to 30 percent) and savings in capital investment (up to 70 percent). For the Leamington greenhouse industry these savings were estimated to be more than \$100 million during the four years from 1993 to 1997. Advances in plant pathology and entomology have resulted in pesticide reduction, which has created a more valuable commodity in a consumer environment that has become concerned with the use of chemicals in agriculture and which has helped the industry to expand in both foreign and domestic markets. The growing interest in organic farming is an indication of this trend. Equally, the shift from hand pollination to the use of bumble bees in the greenhouse tomato industry, which was also a result of Harrow research, has brought about significant savings in labour costs to the grower.

The development and commercial introduction of the Harrow Fertigation Manager (HFM) are among other significant achievements. The HFM is a computer-controlled, multifertilizer injector system for the precise application of fertilizers to any crop that uses preprogramed seasonal fertigation (from "fertilize" and "irrigate") programs (Whitfield and Papadopoulos 1999). The Harrow Greenhouse Crop Manager (HGCM) is a computer program that will help growers "make management decisions on the spot." This project began in 1992 with funding from the federal Program of Energy Research and Development and a grant from the provincial program Food Systems 2002 to design "an expert system to diagnose insect and disease pests of cucumbers and tomatoes." What was originally an attempt at pest control has developed into a significant cost-control program. It has "evolved into a highly interactive decision-support system to help growers improve fruit quality and yield, use energy more efficiently, reduce fertilizer and water waste, and apply alterative control strategies to eliminate pesticide use" ("Greenhouse Growers"). As a result, the program has been a major contributor to cost reduction.

In Ontario the greenhouse industry has also experienced several significant changes. First, there has been a shift from soil to hydroponics. In 1998, 86 percent of greenhouse tomatoes and 95 percent of greenhouse cucumbers were grown in rockwool, not soil (Khosla 1998). Second, there has been a shift in production from a pink variety of tomato to the red beefsteak, which has received much greater consumer acceptance. It is also more disease resistant and higher yielding. Third, packaging has progressed from hand-packing to state-of-the-art automated lines that sort by weight, size, and colour and automatically place appropriate product code stickers on the vegetables. According to Denton Hoffman, "these dramatic changes have helped establish Ontario as a quality player in the North American premium tomato market." He also points out that these changes have turned Ontario growers into leaders in productivity in the North American industry. Productivity has doubled for greenhouse tomatoes and seedless cucumbers in the past decade. New cultivars allow Ontario growers to produce over forty pounds of tomatoes and eighty seedless cucumbers per plant (OGVPMB 1998, 8).

PROFITABILITY

To determine the profitability of Ontario (and, more specifically, Leamington) horticulture and its greenhouse sector, expenses have

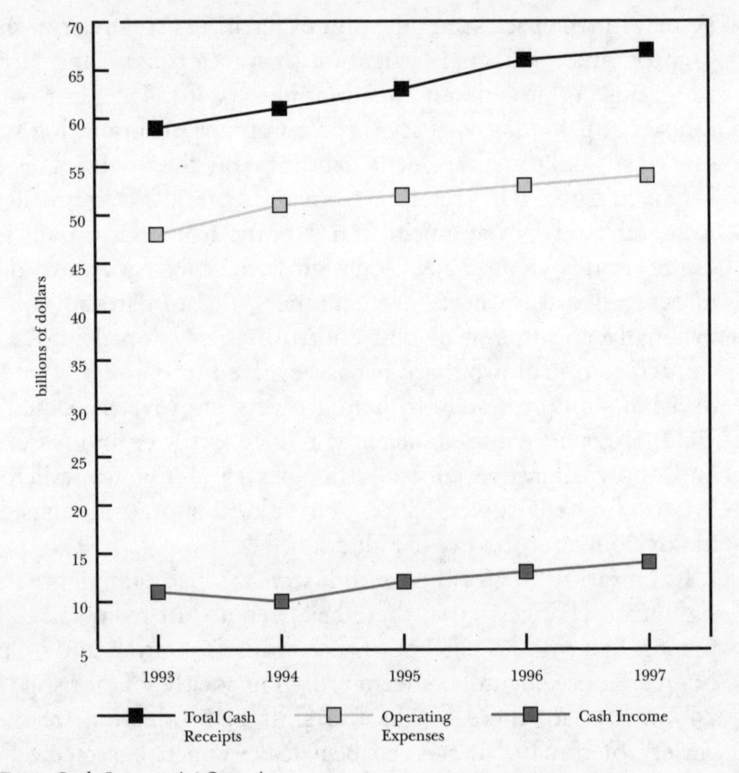

Fig. 4 Farm Cash Income in Ontario, 1993–97

to be deducted from sales (or "farm value," the term used to refer to sales in OMAFRA statistics). As can be seen from figure 4, agricultural cash incomes increased between 1993 and 1997, despite increasing operating expenses. That the incomes were sufficient to keep the industry viable between 1995 and 1997 is demonstrated in table 5.7, from which it can be inferred that incomes were sufficiently greater than expenses to provide opportunities for the capitalization of income into new farm assets. The relative increase in farm assets was greater than the relative increase in costs. New assets were acquired either through the capitalization of income or through debt financing. This may indicate that incomes were sufficient to meet all existing expenses, with the surplus being reinvested in the industry, or it may indicate that financial institutions felt sufficiently optimistic about the industry to lend funds for capital expenditures to expand operations. This assumption is supported by the significant equity held by farmers in the assets of farm

Table 5.7
Ontario Greenhouse Industry, Total Costs vs. Asset Value, in Thousands of Dollars

Year	*Total Costs*	*Percentage Increase*	*Market Value of Assets*	*Percentage Increase*
1995	234,391		600,371	
1996	264,606	13	717,435	19
1997	283,325	7	830,500	16

Source: Statistics Canada, Catalogue No. 22-202.
Note: Costs include total payroll for full-time and part-time labour, fuel, and the purchase value of plants, seeds, flowers, cuttings, and bulbs. Market value includes fair market prices of land, buildings, equipment, and machinery.

Table 5.8
Farmer Equity in Assets of Farm Production in Ontario, in Thousands of Dollars

	1993	*1994*
Total assets	35,902,353	36,170,374
Total liabilities	4,278,693	4,463,525
Equity	31,623,660	31,706,849
Equity as percentage of assets	88	88

Source: OMAFRA 1995, table 17.

production (table 5.8). An equity position of 88 percent of assets for both 1993 and 1994 indicates a particularly healthy financial position. Even though data for the years following 1994 were not available, these numbers indicate that farming in Ontario is a profitable venture.

The health of the Leamington greenhouse industry is also revealed through the economic decisions of the farm operators. It has been estimated that an acre under glass costs between $700,000 and $1 million to build, including the boiler, the computer system, and the greenhouse skeleton. If each acre produces $150,000 worth of tomatoes, it can take several years to recover these investment costs (Welch 2000a). However, these costs have not discouraged the growers. In the last ten years twenty-one of the forty-five growers interviewed have purchased a total of 734 acres of land, averaging 35 acres per grower. Over the same period the forty-five growers have invested $72.6 million in their greenhouse production, averaging $1.4 million per grower. What is particularly interesting is that ten of these farmers used internal financing for their expansions.

Table 5.9
Average Farm Capital, by Region, 1996

Region	*Number of Farms*	*Average Farm Capital*
Ontario	67,520	605,168
Essex	2,109	708,151
Leamington	444	798,856

Table 5.10
Total Gross Farm Receipts, by Region, 1996

Region	*Total Sales*	*Number of Farms*	*Average Sales*
Canada	32,230,356,237	276,548	116,545
Ontario	7,778,476,483	67,520	115,202
Essex	315,742,917	2,109	149,712
Leamington	130,822,715	444	294,645

These investments have brought the average capital value of the greenhouses to $3 million, which is significantly higher than the average value of the farms outside the greenhouse sector. According to the agricultural census of 1996, the capital value of the average farm in the Leamington area was $798,856. The provincial average was even lower (table 5.9). The average annual farm income for the thirty vegetable producers who provided full financial information was $295,774, in spite of the fact that their average mortgage payments were $215,133 annually. When household income derived from greenhouse operations was added to the reported profits, the calculated average was $393,833. Sales averaged $2,929,833 among the thirty growers. If one grower who owns thirty-one-acre greenhouse enterprise and who reported having made sales of $68 million is removed from the calculations, the average sales were $1,513,620, considerably higher than the average sales reported by all Leamington farms in the 1996 census (table 5.10). The difference can be attributed to the higher sales made in the greenhouse sector, as well as to increasing sales in the greenhouse industry since 1996. It is also noteworthy that average provincial and national farm sales were considerably lower than the average sales reported by the Leamington growers in that census year.

Thus, the Leamington greenhouse industry is experiencing healthy growth. This is not to say, however, that all greenhouse vegetable growers have enjoyed it to the same extent. The aggregate statistics mask differences among the greenhouse growers. For instance, among the growers interviewed in the study, the size of

Table 5.11
Distribution of Greenhouses, by Size

Acreage	*Number*
0–4.9	15
5–9.9	18
10–19.9	3
20+	4

Table 5.12
Distribution of Greenhouses, by Area of Land Purchased in the Last Ten Years

Acreage	*Number*
1–19.9	7
20–49.9	8
50–99.9	4
100+	2

Table 5.13
Distribution of Greenhouses by Farm Capital Value, in Millions of Dollars

Value	*Number*
less than 1	8
1–4.9	25
5–9.9	3
More than 10	3

Note: Figures for the value of one farm were not available.

the greenhouses ranged from one to thirty-one acres. The largest number of greenhouses were in the middle range – between five and ten acres (table 5.11). Not all the interviewed growers have been able to expand their farms to the same extent (table 5.12), and, consequently, the capital value of the greenhouses is similarly differentiated (table 5.13). Finally, as many as four growers claimed that they had lost money in the previous year, and the levels of profit varied for the others (table 5.14).

The fact that some greenhouse vegetable growers still feel vulnerable is illustrated by the following remarks by one of them:

And if our son hadn't joined us – he is a teacher and he and his wife are working – he came and we expanded, we were at the point when we should be selling but he didn't want us to, that's why we are still in

Table 5.14
Distribution of Greenhouses by Profit Made in 1998

Profit (in Thousands of Dollars)	*Number*
Loss	4
0–49	10
50–99	2
100–499	15
500–999	3
1,000 +	1
No answer	5

business ... And we went through some really hard times when the interest rates were high in the '80s and a lot of our friends – all my friends were in the same business – a lot of them have folded their farms. They did so because the fuel prices were high. We keep wondering what's going on. We can't understand what's going on, all this building here. There isn't much left over. Just now, they are telling us that the gas prices, how we heat our places, they are going to increase 40 percent. They were just put up twice since last year. And that's why my husband and I are not enthused about all this. We've gone through cycles.

SUMMARY

The Leamington greenhouse industry has enjoyed significant growth over the last decade. Many Leamington greenhouse vegetable producers have been able to make a profit and expand their production. However, not all Leamington growers have experienced growth to the same extent. There are some greenhouses in Leamington that are still small-scale, family-operated businesses. Some producers claim to have lost money, and some have not been able to expand. At the same time, the economic growth of some producers has been impressive. As we have seen, despite the differences in the size of production, farm capital value, and profits, most greenhouse growers depend on offshore workers. Large and successful greenhouse operators complain about their inability to find reliable local workers just as much as, if not more than, small growers. But the explanation that links the inability of the growers to attract local labour to their economic vulnerability can hardly apply to the many farmers who can afford to increase their workers' wages and extend them some benefits. Yet, as discussed in chapter 4, Leamington growers feel that if they were to increase the wages by a few dollars, workers would still be reluctant to work in the greenhouses, particularly in the summer. Furthermore, the evidence in chapter 3 reveals that growers need workers who are not free to take time off work for any excuse – be it a religious celebration, a social event, or sickness – at the time of high demand, when crops are ready to be picked. This requirement for captive labour is unrelated to the salaries paid. Even if agricultural salaries were to increase significantly, local workers would not consent to become chained to the job. It is this need to have workers who are available for work on demand, and not economic vulnerability, that makes most Leamington greenhouse growers dependent on offshore labour.

THIS CH. DEMONSTRATES THAT LEAMINGTON GREENHOUSE FARMERS ARE HIGHLY PROFITABLE. 70% OF THEIR TOMATO PRODUCTION WAS EXPORTED TO THE U.S. IN 1998, SO NO "SELF-SUFFICIENCY" ARGUMENT CAN BE MADE TO PROTECT THESE FARMERS – THEY'RE JUST GREEDY CAPITALISTS. THE TOP 18 GREENHOUSE OPERATIONS IN CANADA CONSTITUTE 0.5% OF ALL CDN. GREENHOUSE PRODUCERS BUT ACCT. FOR 31% OF TOT. CDN. AREA. LEAMINGTON PRODUCERS CONTROL AN AVERAGE AREA SEVEN TIMES LARGER THAN THE CDN. AVG. GREENHOUSE.

BY ALL MEASURES, THESE ARE ENLIGHTENED FARMERS WHO HAVE POOLED THEIR RESOURCES INTO A RESEARCH CENTRE TO IMPROVE THEIR CROPS, GOING W/ ORGANIC TRENDS. INCOME HAS INCREASED MORE THAN OPERATING COSTS, SO PROFITS TOO HAVE INCREASED DURING THE 1990S.

YET, THE FACT REMAINS THAT THESE OPERATIONS REQUIRE UNFREE LABOUR, AVAILABLE TO WORK ON DEMAND, => IT IS NOT THE SECTOR'S VULNERABILITY BUT THIS STRUCTURAL NECESSITY "THAT MAKES MOST LEAMINGTON GREENHOUSE GROWERS DEPEND ON OFFSHORE LABOUR" (85).

[illegible]

[illegible]

PART TWO

Mexican Harvesters

6 From Mexico with Two Hands

One evening after Celia had put their three children to sleep, Rodolfo decided to talk to his wife. He had been thinking about working in Canada for a long time, but until then he had continued to hope that it would not be necessary. Now that his eldest son was graduating from the *primaria** and their second son was to go to school, the decision could not be put off any longer. Rodolfo's earnings as a *jornalero* simply were not enough to pay the expenses required to send their eldest son to the *secundaria*. The cost of books, transportation, and uniforms kept mounting, while there was not enough work for him throughout the year and the salaries paid to *jornaleros* were too low to make ends meet. And then, he thought, how long could they keep on living in one room in the house that his father gave him? What would happen when they had more children? Things would not have been so bad if his wife, Celia, had not needed a caesarian section with her third pregnancy. But now it would take Rodolfo a few years to pay off the debts for the private medical care that she had needed. And then his children had been sick a few times, and that had meant more debts.

Many people from Rodolfo's *rancho* went up north to earn money. Those with a bit of money could afford to pay a *coyote* the fifteen hundred dollars required to take them across the border to the United States. Rodolfo could not even dream of having so

*For Spanish words, see the glossary at the back of this book.

much money now, but there was an alternative. His cousins Secundo and Gerardo had gone to work in Canada. They had come back and started building their own houses right away. Gerardo told Rodolfo that if he wanted, he would take him to the *oficinas* in Mexico City and put in a good word for him with the *licenciado*. Rodolfo had never lived outside his *rancho*, and the thought of going all the way to Canada scared him. But if he was lucky, he would be sent to work at a place where there would be others from his *rancho*. More people had been going to Canada from here. They said it was hard being away from your family for such a long time, but just the thought of how he and Celia could use the money he would bring back meant that nothing else mattered.

Gerardo told Rodolfo that it was not always easy to get selected to go to Canada. There were lots of people in the *Secretaría* interested in going to work there. Sometimes, if the *licenciado* was not convinced that you were poor enough to go, a bribe would help. But Rodolfo was hoping that he would not have to pay it. After all, he had no land and no other way of making a living besides being a *jornalero*. Gerardo told Rodolfo that he had to be persistent and insist that he really needed a chance to go to Canada but that if he was not lucky this year, he should not despair. He had a good chance of being selected the following year, provided that he was healthy. Gerardo also told Rodolfo that he would need to borrow money to be able to travel to Mexico City. First, he would need to fill out an application form; then he would need to take two medical exams; and then he would need to get a passport from the Oficinas de las Exteriores.

Others in his *rancho* cooperated. They got a pick-up truck to take them to Mexico City, but he could not sleep in the street there, and the hotels were expensive. Perhaps Rodolfo could borrow some money from his uncle Aurelio. When he returned from Canada, he would pay him back right away. Rodolfo needed to discuss all this with Celia, but surely she would understand that they could not build a house, send their children to school, pay medical bills, and clothe themselves and their children if he stayed in his *rancho*. It would be hard for Celia to be on her own, and the children needed their father to discipline them, but they had to make a sacrifice in order to better themselves.

Rodolfo's story is a composite. It underscores several themes common to many migrants coming to work in Canada on a seasonal basis.

Most of them are from small villages, where they work as day workers. They are married and have children, but what they earn is insufficient to meet their household needs. They need to build a house and send their children to school – first *primaria* (primary school), then *secundaria* (secondary school), and then *preparatoria* (high school). They are therefore drawn to Canada and need to continue working there for many years. They generally prefer employment in Canada to illegal migration to the United States, partly because of the high fee charged by the middle man who takes migrants over the U.S. border illegally but also because of the security of employment that the Canadian program offers them, among other things.

Migration to Canada is not an automatic process. Temporary agricultural workers are selected by a *licenciado* at the Secretaría de Trabajo y Provisión Social (Ministry of Labour and Social Planning) in Mexico City. Mexican workers refer to this ministry simply as the *Secretaría* or the *oficinas* (offices). It is the same office that determines whether they will continue working in Canada. A worker who satisfies the selection criteria and passes medical examinations is issued a passport by the Oficinas de las Exteriores (the Ministry of External Relations).

POVERTY AND MIGRATION: MIGRANTS' STORIES

In his analysis of U.S.-bound Mexican migration Dussel Peters observes that while it has various underlying causes, unlike migrants from other countries to the United States, Mexicans migrate mainly for economic reasons: the lack of stable and well-paid jobs in Mexico largely explains their migration (Dussel Peters 1998, 56–7). This observation is true of most participants in the Canadian program as well. They come to work in Canada because jobs in Mexico are scarce, insecure, and badly paid; inflation has made it very difficult for them to provide for their households. The following are some comments by Mexican workers explaining why they felt compelled to leave their homes and come to work in Canada.

Rubilio, from Guanajuato, remembered his life eleven years ago, before he started coming to Canada: "I had a brick house with a roof that leaked when it rained. I used to put blankets over it because the roof was rotten. And the kitchen was right there as well. And we did not have a single bed. And that's how we all lived. My wife used to make tortillas outdoors ... It's very sad to be poor." Irineo, from Chiapas, explained that:

There is no work. If there is work, there is very little of it. Those who have it are not going to leave it. Recently the situation has become more critical because of the problems that the Mexican government has had. There are more people and less work.

In Chiapas the situation is very critical. Last year ... there was an uprising against the government for the same economic reasons that have marginalized us ... Those who have land work on it by themselves, and they are paid poorly because the government pays them little. So there are very few people who have money. The *caciques* (land barons) occupy all the land. And for the marginalized people, the poor people, there is no place to work. And if they give them work, they pay them some fifteen or twenty pesos per day. Not enough even for beans. And imagine if one has four or five children, how is he going to feed them?

Gregorio, from Guanajuato, painted a similar picture: "There is always very little work. There used to be work when there were *patrones*, but now there are no *patrones*. There used to be people who gave you work before, but with time these opportunities have disappeared. And the work is badly paid. The majority of people work on their own land and do not hire people. And the poor people have to go to other places to look for work."

One possible way to resolve these problems would be to migrate to the capital. But jobs are scarce there as well. Filogonio commented that "In Mexico City there are a lot of people but not much work ... So many people from other states go to Mexico City in search of work. There is simply not enough for everyone." Hugolino, a forty-two-year-old industrial mechanic from Mexico City, who lost the job he had had for fifteen years, discussed his situation:

I am already old. It's not the same as before. They say that they need young guys who are twenty, twenty-five ... So I kept on trying. I went to different places, learned different occupations. I could no longer find anything as a mechanic. So, "Do you want to be a driver?" "Yes." So I drove. So I worked where I could, as an unskilled worker too, whatever ... The truth is that the political situation that we had since 1995 made us leave. I am a product of this poor administration. Many have come here. I never thought of leaving the country. I had never been apart from my family ...

The government told us that in two, three months things would get better. But two years went by and everything stayed the same. Instead, they closed more businesses, and one goes to look for a job and they tell you that they do not need any more people. Instead, they fire more workers. We are many and that's how problems occur.

POVERTY: UNDERLYING CAUSES

Mexican north-bound migration motivated by a search for more stable and better paid jobs is not a new phenomenon. It dates back to the last century (Jones 1984). Then, as now, the displacement of Mexican rural producers by agribusinesses, many of which were foreign-owned, played an important role in stimulating migratory movements.

The process of land concentration in the hands of large corporations has long, historic roots in Mexico (see Ramirez 1989). While the agrarian reform that followed the Mexican Revolution of 1910 slowed down the process to a certain degree, the development strategies adopted in Mexico in the twentieth century reversed the achievements of the Mexican peasantry and exacerbated the problems of massive rural underemployment and poverty. Among those strategies, the policy of import-substituting industrialization (ISI), which was pursued between 1946 and 1966, played a very significant role in the destruction of Mexico's subsistence agriculture, particularly the production of rice, beans, maize, and wheat by emphasizing industrial development and the provision of cheap food to the industrial working class. While certain agricultural sectors did receive significant boosts through the Green Revolution during the ISI years, production in others declined. The Green Revolution concentrated on high-yielding varieties of rice and wheat that required largely subsidized outlays for irrigation, fertilizer, and pesticides. Assistance went predominantly to commercial agriculture and large estates. While the government subsidized credit, research, mechanization, marketing facilities, and the distribution of implements for the large private farms, *ejidos* (communal land holdings) and private peasant farms received no support. As a result, the economic status of the vast majority of farmers who raised maize and beans declined, and Mexico began to depend heavily on imports of basic grains from the United States (Warnock 1995, 194–5).

Some of these trends were offset by the policies adopted by Presidents Echeverría and López Portillo. President Echeverría (1970–76) introduced a program to help subsistence farmers., which included price support, subsidized credit, discount prices for agricultural inputs, crop insurance, storage and marketing assistance, and education and health services. Additionally, new technologies were provided for peasants depending on rain-fed agriculture. Simultaneously, he introduced some changes to strengthen the *ejido*, or rural community, system and redistribute more land to *ejidos*, although most of the land granted to them was of poor quality (Warnock 1995, 195–6). President López Portillo (1976–82) also attempted to promote self-sufficiency in food. In March 1980, SAM (the Mexican Food System) was introduced to give support to dryland and tropical agriculture, in an attempt to increase the production of basic foods. The objective of SAM was to reach self-sufficiency in the production of corn and beans by providing assistance and investment to small farmers in rain-fed districts. Important new policies included the provision of credit from the National Rural Credit Bank (Banrural) and a government guarantee to share the risk of crop failure with farmers (196–7).

The debt crisis of 1982 put an end to these policies. In 1982 President de la Madrid abandoned the SAM policy and adopted several measures known as "structural adjustment policies," but also as "neo-liberal policies," which were dictated by the International Monetary Fund (IMF). These policies included the following measures: raising interest rates to limit consumer spending and discourage the flight of capital; reducing government spending by removing subsidies on food, fuel, and public transport and cutting social spending on health, housing, and education; devaluing the currency to encourage exports and discourage imports; removing price controls; and privatizing state companies (Green 1995, 42; Dussel Peters 1998, 58–9; Warnock 1995: 197–8). While the policy of price deregulation could have allowed subsistence producers to charge higher prices, in reality many subsistence farmers were driven out of business by the competition from cheap imports made possible through these policies (Green 1995, 99). Under President de la Madrid, credit to agricultural farmers was curtailed, the budget for agriculture was drastically cut, and many research institutes and technical assistance programs were closed (Warnock 1995, 197; Russell 1994, 194).

The structural adjustment policies have had a direct impact on subsistence agriculture. For example, the subsistence sector in Mexico has experienced a significant decline in production that meant that basic foods had to be imported (Russell 1994, 194), even while export agriculture was being rigorously promoted. The promotion of export activities in rural areas is associated with the presence of the transnational food corporations that have been provided with state subsidies for credit, fertilizers, and other farm inputs. Land that had been cultivated for basic food crops such as maize, beans, and wheat is now producing animal feed crops such as sorghum. The change to sorghum has also reduced rural employment requirements, since sorghum production requires lower labour inputs than does maize production. Additionally, subsistence crops have been replaced by luxury export crops like strawberries, tomatoes, melons, grapes, pineapples, and broccoli (Warnock 1995, 199–200; Green 1995, 107–8).

In fact, one of the most significant outcomes of these policies has been the growth of underemployment, particularly in the formal economy. While some economic activities – especially the production of automobiles and auto parts, basic petrochemicals, and glass and electronic goods – experienced growth, they failed to provide adequate employment for the workers expelled from other economic sectors, due to the capital-intensive nature of production in these growth sectors. The economically active population increased by 17.05 million between 1980 and 1996, but employment increased by less than 2 million formal jobs (Dussel Peters 1998, 66). A considerable decline in the share of the labour force employed in agriculture and manufacturing was accompanied by an increase in the number of jobs in the service sector, particularly in community services and construction, areas characterized by particularly low wages (Dussel Peters 1998, 67–8). Warnock (1995, 205) estimates that the average farm labourer in Mexico works between sixty and ninety days a year. The farm labour force in Mexico is comprised of around seven million individuals, of whom 4.5 million are *jornaleros* (day workers) who do not have regular employment but rather who migrate from farm to farm in search of work. During the period discussed, average real wages, as well as real minimum wages, dropped substantially: on average, in 1996 real wages represented only 60 percent of their value in 1980, and the real minimum wage in 1996 was only 27 percent of the value in 1980 (Dussel Peters 1998, 69–70).

This pattern of economic development affects migration in two ways. First, by generating landlessness, underemployment, and unemployment, it creates masses of people desperate enough to undertake a journey up north. At the same time, this development process also stimulates expansion of economic activities in some areas, creating a need for capital. Some researchers have argued that it is not the poorest workers who migrate but those originating from areas undergoing modernization. They do so in order to obtain capital to invest in productive activities (Massey and Espinosa 1997, 953–4). But in the case of Canada-bound migration, because of the state-controlled nature of the migration process discussed below, it is the poorest workers who dominate the flow.

WORK IN CANADA AND THE UNITED STATES COMPARED

For Mexican migrants, legal work in Canada is preferable to illegal work in the United States. Of the 565 individuals interviewed in the present study, 158 had worked in the United States *de mojados* (as "wetbacks," or illegally) having been detained and deported or having eventually left voluntarily. Those who had not worked illegally in the United States gave various reasons for not doing so. Some were discouraged by the high costs of getting into the United States (as mentioned, a *coyote* charges $U.S. 1,500 to take illegal migrants over the border today), others by the uncertainty of finding work in the illegal labour market. Some stated that the living conditions that were related to their illegal status resulted in a precarious situation and that the need to pay rent and the presence of temptations such as bars, movies, and dances absorbed much of their earnings and reduced what could be sent home to their families in Mexico, thereby obviously also reducing the benefits of working *de mojados*. The workers were in general agreement that although earnings may be higher in the United States and although the U.S. dollar had greater purchasing power, the net benefit was low, because they were unable to save much of their earings while they were in the United States.

In Canada, on the other hand, Mexicans work long hours but are able to save a significant portion of their earnings, which enabled them to remit greater amounts to their families in Mexico. This situation may be attributed to the fact that working long hours leaves little time for entertainment, as well as to the fact that many workers are relatively isolated on the farms, with little opportunity to visit the cities where their wages could be spent in bars and restaurants: they may have little choice about remitting their earnings to Mexico. Colby (1997, 27) estimates that remittances sent from migrants in Canada are significantly higher than remittances from migrants in the United States. Her study of 1994–95 of migrant workers from Oaxaca found the average monthly remittances from Canadian migrants to be a thousand dollars, compared to two hundred dollars per month remitted by a migrant from the United States. Furthermore, as mentioned, many participants in the Canadian program value the job and wage security they enjoy in Canada (Colby 1997, 19; Argüello 1993, 96).

The following comments on how work in the United States compares to work in Canada were made by Mexican workers interviewed in this study who had experienced both and by others who had not:

- Here we have a contract. There we work sometimes, and other times we don't.
- Here we have a house to come back to.
- There you work from Monday to Friday, but on Saturday and Sunday you have to go out, and these are extra expenses. Here we can save better.
- When one goes to the U.S., one has to stay there for many years, being separated from one's family.
- For me it would be difficult, and for my wife and my children as well, because they would not know anything about me, because one never knows where there will be a job. But here, on the other hand, we come on a contract. And from here I can keep in touch with them by phone and they know my address and in case I cannot write to them or my letters get lost, they can go to External Relations in Mexico and get our addresses.

One worker said that he would not work in a racist country such as the United States. He believed that Mexicans in Canada did not experience as much racism. Argüello (1993, 96) reports that Canada-bound Mexican migrants interviewed in his study prefer employment in Canada because they can "live like people" (*vivir como la gente*) in Canada, enjoying better nutrition and access to certain comforts, like a bathroom.

Not only do most men chose to work in Canada rather than in the United States, but their wives prefer it when their husbands are employed in Canada. Colby (1997, 26) lists several reasons why most women prefer their husbands to go to Canada. Among them are the following:

- There they don't drink and get in trouble with the police.
- My husband sends money home. When he was in the U.S. he forgot to send money and forgot about this town.
- We know he will come home each year by September, or at least by the Day of the Dead.
- In Canada he doesn't have girlfriends or other babies.
- To go to Canada he doesn't have to suffer the dangers and expense of crossing the border to the U.S.
- My children need their father. When he goes to Canada I know he will be back. When he went to the U.S., I never knew where he was, what he was doing, if he had a job, if he was sick or hurt, or ever coming home.

THE SECRETARÍA

In selecting workers for the Seasonal Agricultural Workers Program, the Mexican Ministry of Labour and Social Planning follows set guidelines. In order to qualify for the program a worker must meet a specific set of criteria. To satisfy the requirements of Canadian employers, Mexican applicants must have work experience in agriculture. Since the main objective of the program from the Mexican perspective is to assist those who are most in need, applicants with the lowest levels of education, applicants who are landless, or applicants who lack other means of subsistence (e.g., a business) are given preference to participate in the program. Workers who are married with children are given preference over single or childless workers, to ensure that the workers return to Mexico at the end of their employment term. This criterion satisfies Canadian immigration authorities that this seasonal migration will not become permanent immigration. While the statement of understanding signed by representatives of the Canadian and Mexican governments sets the minimum age for participating in the program at eighteen, the Ministry of Labour has chosen to raise the minimum age to twenty-five, in order to select candidates who are more likely to have large families and who are therefore deemed more deserving of participating and more likely to return to Mexico when the term of employment is over.

All participants in the program are required to have completed military service in Mexico. Very few women participate, because Canadian farmers want to avoid the cost of supplying separate living quarters for women. Therefore, though gender is not an actual criterion for participation, mostly men are requested by farmers in Canada. In practice, the guidelines for the program are not strictly adhered to. Among those interviewed for this study, some were single, some had relatively high levels of education, and some were land or business owners.

In Canada employers are allowed to request specific workers. For example, in 1996, 70 percent of the 4,481 Mexicans participating in the program were specified or named by their employers (FARMS 1997). Participants who are known to growers and have proven to be good workers are obviously an asset. As a result, many workers return to the same farm for several years. The average number of years spent working in Canada was 6.5 for the workers interviewed in the Leamington area. The tendency of Mexican seasonal workers to return to work in Canada, if not to the same farm, is akin to the behaviour of Caribbean workers employed near London, Ontario (see Cecil and Ebanks 1992), and of Caribbean cane-cutters in the United States (see Wood and McCoy 1985, 273).

Upon their return from Canada, workers have to report to the *oficinas*. They are required to provide an account of their expenses to the *licenciado* who will want to make sure that the Canadian-earned money is not spent frivolously by the workers and their families. The program participants also hand in an evaluation of their work in Canada that every employer is requested to write. If the evaluation is positive, the worker expects to be invited to reapply.

The bureaucratic process through which the Ministry of Labour selects participants in the Canadian Seasonal Agricultural Workers Program contrasts with a considerably less orderly recruitment procedure used in the United States during the *bracero* years (1942–64). At that time aspiring *braceros* often had to pay a *mordida* (bribe) to obtain a permit from municipal Mexican officials that allowed workers to go to central recruiting centres, where there may have been as many as ten workers for each *bracero* vacancy. Workers often paid another bribe at these centres, in order to get a preferential opportunity to apply for the limited vacancies. From the labour pools at these centres, the Department of Labour officials selected *braceros* to be sent to the border reception centres where contracts were signed (Calavita 1992, 62–3). Growers often participated directly in the hiring process, arriving at contracting centres, choosing the men personally, and then negotiating terms and conditions of employment with the applicants themselves (Galarza 1964, 83).

CHARACTERISTICS OF THE PROGRAM PARTICIPANTS

Most Mexicans participating in the Canadian program come from the Mexican states located around the capital. Because workers must travel to the Ministry of Labour in Mexico City to apply to the program, in 1996 more than two-thirds of the program participants (69.2 percent) were from four Mexican states – Tlaxcala (23.2 percent), Guanajuato (18.5 percent), Mexico (17.5 percent), and Hidalgo (10.5 percent) – all located in central Mexico and all relatively close to Mexico City. While relative proximity to the agency administering the program may be significant, social networks may be of even greater importance in this skewed geographic distribution. The importance of social networks is demonstrated by the fact that in the state of Guanajuato, virtually all program participants come from seven municipalities – Irapuato, Abasolo, Acámbaro, Pénjamo, Jaral del Progreso, Jerécuaro, and Salvatierra. The male population in these seven municipalities comprises only 27 percent of the male population of the state of Guanajuato (table 6.1). The lion's share of program participants come from the municipality of Irapuato, the male population of which comprises only 9.4 percent of the male population of the state. With the exception of Jerécuaro, each of these municipalities sends a high concentration of migrants. For example, almost three-quarters of Canada-bound migrants come from twelve communities in Irapuato, with 259 workers coming from the single community of San Cristóbal, which has a total male population of 1,624 people, according to the population census of 1996. Two communities in Acámbaro, Los Desmontes

Table 6.1
Male Population and the Number of Program Participants for Seven Selected Municipalities, State of Guanajuato, 1996

Municipality	*Male Population*	*Percentage Participating*	*Number of Participants*
Abasolo	36,057	1.7	79
Acámbaro	53,370	2.5	189
Irapuato	200,030	9.4	966
Jaral Progreso	15,036	7.0	44
Jerécuaro	25,769	1.2	309
Pénjamo	67,401	3.2	23
Salvatierra	46,103	2.2	130
Total, Guanajuato	2,139,104	100.0	1,740

Source: Mexican Ministry of Labour and Social Planning; 1996 Population Census.

and San Juan El Viejo, send more than half the migrants from the area, while more than half the migrants in Salvatierra come from a single community, San Pedro de los Naranjos. In 1996 municipalities with less than a hundred Canada-bound migrants, communities such as Abasolo, Jaral del Progreso, and Pénjamo, sent a concentration of migrants from only one or two communities (table 6.2).

In the state of Tlaxcala the total of 2,023 participants came from more than eighty different municipalities in 1998. However, five municipalities – Domingo Arenas, Ixtlacuixtla de Mariano Matamo, Lazaro Cárdenas, Mariano Arista, and San Lucas Tecopilco – provided 866 program participants, or 42.8 percent of the total for the state. Furthermore, within four of these municipalities participants were concentrated in particular communities (table 6.3). An examination of the lists of program participants acquired from the Ministry of Labour suggests that kinship may play a certain role in this migration. Numerous people with the same paternal and maternal surnames and those with either the same paternal or maternal surnames are easily detected on the list, although they are not always related by blood.

Because of the criteria used by the Ministry of Labour to select program participants, workers migrating to Canada differ from migrants to the United States (legal and, particularly, illegal migrants) in several ways. Various studies conducted in the United States point out that many migrants come from the western Mexican

Table 6.2
Communities of Origin of Canada-Bound Seasonal Migrants from the State of Guanajuato, 1996

Municipality/Community	*Number*	*Percentage*
Abasolo	79	100.0
Horta	20	25.3
Nuevo de la Cruz	39	49.4
Acámbaro	189	100.0
Los Desmontes	58	30.7
San Juan El Viejo	40	21.2
Monte Prieto	16	8.5
Tocuaro	12	6.3
Irapuato	966	100.0
San Cristóbal	259	26.8
Cuchicuato	77	8.0
Guadelupe de Rivera[1]	58	6.0
La Soledad	52	5.4
San Javier	50	5.2
Providencia de Perez	48	5.0
Venado de Yostiro	35	3.6
Nuevo de Dolores	31	3.2
Sn Luis de Janamo	28	2.9
El Coecillo	25	2.6
Purisima de Covarrubias	24	2.5
San Jose de Bernalejo	22	2.3
Jaral del Progreso	44	100.0
Santiago Capitiro	29	65.0
El Tecolote	5	11.4
Pénjamo	23	100.0
La Troja	14	60.8
Zapote de Barajas	4	17.3
Salvatierra	130	100.0
San Pedro de los Naranjos	66	50.8
La Magdalena	21	16.1
Maravitio del Encinal	8	6.1
Aquiles Serdan	7	5.4

Source: Calculations are based on the lists printed by the Mexican Ministry of Labour and Social Planning.

[1] Represents the number of migrants from two adjoining communities – Guadelupe de Rivera and Rivera de Guadelupe, with 41 and 17 migrants respectively.

states of Jalisco, Michoacan, and Zacatecas. Most of them are men between the ages of twenty and twenty-eight who lack legal permission to work in the United States. Approximately half are married. Most have higher educational levels than the Mexican national average. These migrants are not among the poorest in their communities, and, in fact, many belong to families with medium to low

Table 6.3
Communities of Origin of Program Participants from the State of Tlaxcala, 1998

Community	*Number*	*Percentage*
Domingo Arenas	199	100
San José Cuamatzingo	72	36
Muños	62	31
Guadelupe Cuanuhtemoc	33	17
Ixtlacuixtla de Mariano Matamo	252	100
San Antonio Atotonilco	124	49
Alpotzonga	56	22
Lazaro Cárdenas	149	100
Sanctorum	116	78
Mariano Arista	149	100
Nanacamilpa	110	74

Source: Mexican Ministry of Labour and Social Planning.

income. While they come mainly from rural and semiurban areas, in the 1980s a greater number than previously had come from medium-sized and large cities (Alejandre, Arias, and Varela 1991, 49–50).

As mentioned, most Canada-bound migrants are from the central Mexican states of Tlaxcala, Guanajuato, Mexico, and Hidalgo. Like their U.S.-bound compatriots, they are males, but they are generally older, because the minimum-age requirement for Canadian participants is set at twenty-five. The average age of the 565 Mexican workers interviewed for this study was thirty-eight. All of them had a legal permit to work in Canada, and most were married. Most also had children, the average being four children per household.

Similarly, differences in the nature of the migratory processes between the two countries are reflected in the differences in educational achievements of the two groups. While those with higher levels of education are more likely to undertake a trip to the United States – a new and a significantly different country where their survival depends on their entrepreneurship – the Canadian program tends to favour workers who have not attained higher educational levels. Generally, workers with secondary or higher levels of education are disqualified from the program, though exceptions are made. Relatively few of the workers interviewed had completed secondary education. Slightly over one-third had no education at all or incomplete primary education, and almost as many had completed primary education only.

The class origin of the workers in Canada and the United States differs as well. Migration to the United States requires financial resources that are not available to very poor people. As Alejandre, Arias, and Varela (1991, 53) point out, those who are most able to pay for the trip migrate to the United States. Others, who are without the necessary financial resources, move to the closest cities within Mexico.* This is not a factor in the Canada-bound migration, however, since the travel expenses of the program are partially covered by the employer and partially deducted from the workers' pay. There are still several expenses associated with participation in the program, however, such as numerous trips to the capital and medical examinations. They are usually covered with the help of private loans, which are repaid upon return from Canada. Moreover, because of the specific criteria of the program, many workers employed in Canada had been extremely poor in Mexico before they started participating in the program. Very few had owned land or other means of making a living, and most were from *ranchos* (small villages). For example, of the participants who were interviewed in the Leamington area, 79 (or slightly more than half) had worked in agriculture in their home communities in Mexico. They did most of this agricultural work as *jornaleros*, or day workers. Twenty-nine of these workers reported having taken both agricultural and nonagricultural jobs, such as masonry. Workers interviewed in rural communities in Mexico were even more likely to have held agricultural jobs. In San Cristóbal, for instance, 71 of the 100 migrants interviewed had worked exclusively in agriculture, while 18 had worked in both agricultural and nonagricultural jobs. And among the 311 workers interviewed in eleven rural communities at the last stage of the research project, 74 percent had agricultural jobs and 8 percent had combined agricultural and nonagricultural jobs.

*Durand and Massey have pointed out that poorer members of the community may join the international movement once the earlier-wave migrants settle down and offer assistance to their friends and relatives, thereby reducing their travel and settlement costs (1992, 17).

SUMMARY

Participants in the Mexican Seasonal Agricultural Workers Program are selected by the Mexican Ministry of Labour and Social Planning from applicants with agricultural backgrounds who need social assistance the most. Consequently, they are drawn predominantly from small villages. Most are landless and poorly educated. They have been agricultural day-workers, with a history of irregular employment and salaries too low to cover medical expenses, daily maintenance, and the educational needs of their children. Of course, there is no shortage of Mexicans who meet these criteria. But only some five thousand participants are selected every year, most of whom are repeat migrants. Those who live in relative proximity to the capital are more likely to undertake a journey to apply for the program. But many poor Mexicans are unaware of the possibility of working in Canada. Because the Ministry of Labour does not advertise the program but relies on social networks to spread the word, most program participants are drawn from a limited number of Mexican villages within four central Mexican states.

THERE IS AN OBVIOUSLY PATERNALISTIC ATTITUDE ON THE PART OF THE MX GVMT IN SELECTING WORKERS BOUND FOR CANADA, AS A SOCIAL-ASSISTANCE CRITERION SEEMS TO PLAY A BIG ROLE. YET, THIS VERY CRITERION, TRANSLATED INTO WORKERS W/ THE GREATEST ECON. NEED, MARRIED W/ CHILDREN, AND MOST OF THEM W/ AG BACKGROUND, WORKS IN FAVOUR OF MEETING THE STRUCTURAL NEEDS OF GREENHOUSE GROWERS: AN UNFREE LABOR FORCE THAT IS WILLING TO DO ANYTHING TO STAY IN GOOD TERMS W/ THE EMPLOYER. "MOST ARE LANDLESS AND POORLY EDUCATED" (105).

7 Captive Labour

The first few weeks in Canada were tough for Rodolfo. Everything was unfamiliar, and he felt lost. He did not know how to do anything – mail a letter to Mexico, open a bank account, place a phone call to his family, or send money to them. And he missed them all so badly. Then he got used to being away from his family. Not that he did not miss them, but he was so busy working that he did not have time to feel lonely.

This year he worked in a *grinado*. This was a word Mexicans in Canada used to refer to a greenhouse. At times it was so hot inside the *grinado* that his shirt was soaking wet by midday. He changed his shirt during his lunch break, and in the evening he washed both shirts and also cooked his meals. After dinner he and the other guys who lived in his house watched television. Sometimes they rented videos in Spanish from *El Cubano*, who came with his van to sell them food, CDs, and phone cards. Sometimes Rodolfo's *paisanos*, those from his region, would come to visit him. His *patrón* did not mind if they did, but Rodolfo heard of other *patrones* who would not allow their workers to have visitors. Every second Sunday a local priest performed a mass in Spanish, and on Sunday afternoons Rodolfo would go to watch a soccer game at the field by the church. For Mexican Independence Day they had a *fiesta* in a huge hall and invited *mariachis* to perform. In general, life was uneventful.

For Rodolfo the last two months were the worst. By then most of the guys felt tired and lonely. They could not wait to go home; they did not care even if they were sent home earlier than the date specified on the forms they received from the *Secretaría*. They got angry at each other at the slightest provocation, and some did not speak to each other at all. After seven months of sharing each others' company at home and at work they had had enough. And then it started to get cold. The first time it snowed Rodolfo was excited. He had never seen snow before. He pulled out his camera and took a few shots of the huge banks. But it was also cold at night. They did not want to use the electric heater that their *patrón* had given them for fear of starting a fire. Rodolfo was looking forward to going home.

Rodolfo worked very hard. Some guys on his farm worked very quickly to impress the *patrón*. At times it was difficult to keep up with them, but Rodolfo had to try hard, or else his *patrón* would not ask for him again. Rodolfo needed badly to come back to work in Canada in the future. And that is why he did not ask for a day off to see a physician when he sprained his ankle. He just kept on working.

In July and August his group worked seven days a week, from seven A.M. until eight P.M., with a one-hour break for lunch. When they had finished harvesting they had to clean the greenhouse. But they did not have to work as hard anymore – only about fifty hours per week. Rodolfo was glad to earn so much money, although he was rather tired. He was glad that he was going home soon.

Mexican workers brought to work in Ontario agriculture fill the gap left by Canadian workers. Not only does the Seasonal Agricultural Workers Program supply labour to Ontario farmers, but it supplies labour that is reliable and docile. Unlike local workers, Mexicans are willing to accept minimum wages for work that is back-breaking, monotonous, and detrimental to their health. Even though Mexican labour is relatively costly because of the high transportation and accommodation costs, for many growers it is extremely valuable because it is unfree. Most Mexican workers stay with the same employer as long as there is work for them to do; they are available to work long hours every day; and they do not take time off work, even when they are sick or injured.

THE PRICE OF MEXICAN LABOUR

Bonacich defines the "price of labour" as the total cost of labour to the employer, including, in addition to wages, the cost of recruitment, transportation, room and board, education, and health care – and the cost of labour unrest (1972, 549). Some of these costs are lower for Mexican workers than domestic labour, and others are higher.

Wages

With respect to the wages paid to Caribbean cane-cutters employed in Florida, Wood and McCoy (1985) point out that even though the "adverse effect wage" set specifically for offshore seasonal workers by the Department of Labour is higher than the legal minimum wage, it is nevertheless lower than the wage domestic workers would accept for the work, which is dirty and arduous. Similarly, even though in Ontario Mexican workers receive 5¢ more than the legal minimum hourly wage, taking into consideration the kind of work they do, the wage is rather low. Very few Canadian residents would agree to work for $6.90 per hour in agriculture. According to Leamington greenhouse growers, the prevailing wage for Canadian workers is between $7.50 and $9.00 per hour (Hill 1999b; Schmidt 1999). Of the thirty-nine growers interviewed in this study, only five paid their local workers $6.90 per hour, with most others stating that they pay local labour between $7.00 and $14.00 per hour. Those growers who paid $8.00 per hour or more tended to have a stable labour force. One grower commented: "They say, 'You've got to hire local people.' And we say, 'Bring them on,' and 'How much do you want us to pay? There is only so much I can pay. Do you want me to pay $8.00? Send them over.' Even at $8.00 there were 200 to 250 spots not filled last year ... There will be orders for 250 to 300 people and even $8.00 orders don't get sold. At $7.00 you won't get anybody coming to your place." Even though the agreement between Canada and Mexico stipulates that workers should be paid the prevailing wage if it is higher than the minimum wage, Mexican wages are lower than the prevailing wage.

In addition to the wage, Mexican harvesters covered under the Employment Standards Act are entitled to receive paid public holidays and vacation pay after having been employed for thirteen weeks. In practice very few do. No Mexican interviewed in Leamington had ever received a paid public holiday, and vacation pay is inconsistently offered, with some workers receiving 4 percent vacation pay, others receiving 2 percent, and some receiving nothing at all. Of the eighty-five Mexican workers interviewed in San Cristóbal who had worked in Ontario, twenty-nine had received some vacation pay, fifty-one had never received it, and five had received it on some farms but not on others. As mentioned earlier in this book, the distinction made by the Employment Standards Act between farm workers and harvesters makes it possible for the growers to avoid paying these benefits to the Mexican workers. Only harvesters are entitled to the benefits, but in order to establish whether Mexican workers have worked for thirteen weeks as harvesters, and not as a combination of harvesters and farm workers, growers need to keep track of all the tasks assigned to these workers. No grower in Leamington seems to bother to do so. Unfortunately, there is no information on whether domestic workers receive these payments. It is possible that with respect to the benefits paid to farm workers, Mexicans are as cheap as domestic workers.

Recruitment

Recruitment of Mexican workers is handled by the Mexican Ministry of Labour, with some assistance from FARMS. While growers save on the cost of advertising in Canada, they pay an administration fee to FARMS calculated at the rate of thirty-five dollars per contract worker. While the dollar cost of this recruitment mechanism may be high by comparison to the cost of hiring domestic workers through the unemployment office of the social services department, the efficient manner in which this recruitment mechanism delivers the required number of workers to the jobs is of extreme importance to the growers, as research in other countries has also shown (see Wood and McCoy 1985, 136–7, for instance).

Contract hiring is in some respects akin to network recruitment. As has been pointed out, in many situations involving the employment of foreign workers, social networks provide an ideal mechanism for recruitment by offering an almost unlimited supply of eager new immigrant applicants. Job vacancies can be filled almost immediately without a need for advertisements (Cornelius 1998, 125–6).

Another advantage of network hiring is outlined by Waldinger (1997, 6–11), who suggests that it allows employers to attract applicants who are already "pre-screened". Once hired, these workers are kept in line by those who have referred them and by others from the same community. Like network recruitment, the Mexican Seasonal Agricultural Workers Program efficiently provides workers who are accountable to those recommending them. As one Leamington greenhouse grower commented, "They are very well screened in Mexico, and when they come here they are certainly very interested in working. They take a lot of pride in what they do here."

Workers' Resistance

With respect to the cost of labour unrest, as discussed in chapter 4, no farm workers in Ontario – whether foreign or local – are allowed to organize. In this sense, offshore workers are just as cheap as domestic workers. But Mexican workers are still more docile than many domestic workers. As one Leamington greenhouse grower comments, "They [Mexicans] don't talk back. They don't tell you what you can do with yourself. A lot of the Canadians tell you, 'You can just shove it.'" One grower tells a story of worker from Quebec who was insubordinate:

> In one case I had a person from Quebec, and he lived together with "offshore" and he told me that he knew the law ... And he says, "It's been proven that you have to give a break, 'cause people get more efficient." In the meantime the rest of the workers have already told me that he was telling them to slow down. He was telling the offshore to slow down. So when they told me that I went beserk. He was a worker from Quebec. He had no job. The government sent him down here. And so he told me, "It's been proven that if you give the guys a break it will be more efficient." I said, "How long of a break do you want?" He said, "twenty minutes." I told him, "If you want to be more efficient, make sure you make the rows like the other guys." He said, "I can't do that." "Why?" "'Cause you're not paying me enough." I said, "Didn't you know when you came here that these were the wages?" "Yeh." "Then why did you come here?" Then he told me I'm crazy. That was the end of him.

While some Mexican workers may complain to their friends (and researchers) of being tired, most nevertheless accept these conditions until they are sent back home. Mexican workers generally do not dare to talk to their *patrones* in this manner. Fernando complained to me privately: "We work like slaves here and no one appreciates our work. They don't even buy us beer to thank us for the back-breaking work that we do for them. They just kick us in the butt and send us back to Mexico. And what do we get? Back problems. We get up at five A.M. and work all day. We get wasted fast here. For each year we live here, we lose two years of our lives. And no one appreciates it." Nevertheless Fernando did not confront his *patrón* and continued working until the work was done. At one cannery, female workers told a journalist of a coworker who asserted herself after having been unfairly blamed for the repeated breakdown of a machine. Not only was she sent back home, but she was also forced to pay for her plane ticket (Welch 2000b).

When Mexican workers engage in any form of protest, including making complaints against their *patrones* to appropriate officials at the Mexican consulate, they may experience severe repercussions, as the following two stories from interviewees illustrate:

Filipe worked on a farm in Quebec. One of his fellow workers got fired, but rather than returning home, he hid in the forest. The police arrived and the situation got tense. In solidarity with the rebellious worker, other Mexicans stayed off work. Finally, the fugitive turned himself in, and his supporters went back to work. The following year, all of them were told that they had been penalized, and if it had not been for Freddy's insistence and an appeal to higher authorities, he would not have been able to return to Canada that year.

One season Anastacio was sent to a farm in Quebec. He felt that Mexican workers were treated like slaves there. They lived in a trailer which was poorly equipped. Finally, not being able to tolerate the deplorable working and living conditions, they demanded improvements from the *patrón*. The latter called the consulate; a consular representative arrived but refused to help the Mexicans. Next year Arturo was told at the Ministry of Labour that he had been penalized. It was only two years after the incident that he was able to return to Canada.

As the second story illustrates, complaining to their representatives at the Mexican consulate does not help the workers. In the past, some farm owners did not allow their workers to call the

Mexican consulate or got angry with them if they did. Today most Mexican workers know that they can call the consulate if they a need to, and a toll-free number is provided for their convenience. However, while the consulate assists them with their personal problems (such as when an illness or death in the family impels a worker to return to Mexico), it is unwilling to interfere when workers experience problems with their *patrones*. In the opinion of the Mexican workers interviewed for this study, the representatives either do nothing useful or they take the side of the farm owners against the interests of the Mexican workers. The last thing official representatives seem to want to do is to get the farm owners into trouble. One worker in Colby's study commented, "the Mexican Consulate can't help us even if they wanted to, because they are at the mercy of the Canadian government and don't want Mexicans to be replaced by [workers from] other countries" (1997, 17). Other workers in Colby's study blamed the problem on the corruption of the Mexican consulate. Similarly, some of the Mexican workers I interviewed in Leamington thought that at times farm owners accused of having violated some rights of the workers paid bribes to Mexican officials, so that their violations would not be reported to Canadian authorities. Preibisch (1998, 9) also cites one Mexican seasonal migrant who said: "I don't know if consuls are bought. If we want to talk to them on the phone, they say, 'I don't have time to see you, behave yourselves,' and that's it" (my translation). The same worker elaborated, "The Ministry comes and asks us, 'How are you?' And we say, 'Well,' because we know that they take the side of the *patrón* ... I tell you that the minute one complains, the *patrón* does not ask for you. And then the ministry comes here, 'And why doesn't your *patrón* want you any more? Perhaps, you didn't work well.' And when we explain to them why, they don't believe you. It's better to put up with it" (my translation).

The handling of Inocente's claim for compensation by the Mexican consulate and my experience with the problem provides a good illustration of its position.

Inocente came to work on a tobacco farm in August of 1996. On the next day, he got a skin rash. The farmer's wife took him to a hospital. The physician mentioned that this was an allergic reaction to the pesticide used on tobacco plants and suggested that, if possible, Inocente should not work in the areas sprayed with this pesticide. Not only did Inocente continue working in the same environment, but the farmer's son applied the pesticide while the workers were picking tobacco not ten metres away. So the rash and the itching continued and spread all over Inocente's chest and back. Some neurological problems followed. Inocente lost sleep and exhibited frequent anxiety attacks. His arms swelled. He was taken to the hospital three more times. For fear of losing his job, he minimized the problems, reporting only a rash on his arms. At the end of September, when he was still rather sick, the farm owner sent him back to Mexico, cutting his contract short by two-and-a-half months.

Having returned to Mexico, Inocente went to the Ministry of Labour to report his situation. He was referred to the Ministry of External Affairs, which contacted the Mexican consulate in Toronto and requested that a compensation claim be initiated on Inocente's behalf. And there the case stalled until Rick, a Windsor labour activist, got to work on it. When Rick first called the Mexican consulate, the person in charge denied any knowledge of the case. Later he admitted having possession of all the relevant papers but claimed that the worker's contract had expired at the end of September, even though the worker had received a form from the Mexican Ministry of Labour stating that he had been contracted until the beginning of December.

When Rick lost interest in Inocente's case, I picked up where he had left off by pursuing Inocente's claim for compensation. Eventually, having consulted the employer and the Mexican consulate, the compensation board made a decision to compensate Inocente for the medical expenses he had incurred in Canada and in Mexico, but not to compensate him for the lost time. The employer argued that Inocente was sent back home two and a half months earlier simply because there was no more work for him to do. A few other Mexican workers from the same farm were sent home at the same time. The interpretation of both the employer and the Mexican consulate was that in the absence of a signed contract, a letter that Mexican workers receive from the Ministry of Labour stipulating the length of their employment does not guarantee that their job in Canada will, in fact, last that long. When there is no work left, employers are free to discharge their workers without penalty.

Then I contacted a Windsor lawyer who felt that Inocente should receive compensation, since the letter from the Ministry of Labour indicating the length of his employment could be interpreted as a promise that was broken by the early termination. The lawyer sent a letter to the former employer outlining his arguments. A few weeks later Inocente called me. He sounded alarmed. He had received a phone call from the Mexican consulate in Toronto. The farm owner's wife had replied to the lawyer and sent a copy of her reply to the consulate. The consular representative who called Inocente was very displeased with him for having contacted a lawyer and convinced Inocente to drop the case.

Given the position of the Mexican consulate vis-à-vis the workers they officially represent, it is no wonder that very few workers make official complaints about their *patrones* to the consulate.

Mexican workers are akin to many workers throughout the world who are not allowed to challenge the exploitative conditions under which they are employed through collective bargaining and the use of strike. However, many among them may engage in "everyday forms of resistance" (Scott 1985). Some "hidden" forms of protest include such actions as slow-downs, "careless" work, theft, and self-inflicted accidents and sickness (Cohen 1987, 181, 200–6).

In the case of Mexican seasonal farm workers, however, even hidden forms of resistance are difficult to sustain. When some workers attempt to slow down, others are not willing to support them for fear of losing their jobs or their chances of returning. While occasionally workers do bargain with their *patrones* or with their supervisors over the speed of work, in most cases they are reluctant to do so for fear of repercussions. The only hidden form of resistance worth mentioning is the exchange of produce by the workers. They are generally allowed by their *patrones* to bring home for their own consumption vegetables and fruit that they harvest on the farm. But in order to save on food costs (and thus indirectly increase their pay) workers from different farms often exchange the produce they bring home. While the symbolic significance of these exchanges should not be underestimated, their financial value is rather low.

Transportation and Housing

The need to cover their transportation and housing costs makes Mexican workers more costly to Ontario growers than domestic workers. Ontario growers have often had to pay to transport Canadian workers from other communities or from out of province. But by comparison with local bus fares, the transportation costs of Mexican workers are rather high. In 1998 the cost of airfare for Mexican workers was $697 and an additional $150 had to be paid for their visas. Of the total cost of $847 per worker, growers could recover $575 by making regular deductions from the workers' pay cheques. In addition, employers were responsible for the costs of ground transportation from the Toronto airport to the farms (FARMS 1999).

Growers were also obliged to provide furnished houses for their workers. Some houses had a washer and dryer and an air conditioner, and they all had a fridge and at least one stove. The cost of maintaining these appliances was an additional expense. The agreement for the employment of seasonal agricultural workers requires employers to provide their workers with living quarters approved by the Ministry of Health, even though it is the employers' responsibility to arrange for an annual inspection of the housing accommodations. Because the Ministry of Health does not inspect housing conditions without an employer's invitation, unless there is a formal complaint, some farmers provide substandard housing (Colby 1997, 17; Smart 1997, 148). But whether decent or dismal, housing is an extra expense for the growers who hire labour from outside the area. By and large, local workers do not need housing, and from this point of view also, offshore workers are more costly to the growers. As one local farmer put it, "I wouldn't bring in offshore if I could get enough people. I have to pay their plane ticket. I have to give them housing. I have to ride them into town. It's not cheap. It's expensive."

We can draw a comparison with Caribbean workers hired through the same program. Satzewich calculated, for instance, that the monetary costs of Caribbean labour were higher than the costs of local labour. In 1966 Jamaican seasonal workers were paid on average $1.34 per hour. When contributions to the Canada Pension Plan, the costs of accommodation, and the costs of transportation, recreation, and foodstuffs were added, their average salary increased to $1.81 per hour, which was 50¢ higher than the hourly salary paid to domestic labour when the contribution to the Canada Pension Plan was added to it (1991, 112–13).

COST-REDUCING BENEFITS

Satzewich argues that the differences in wages between Caribbean and domestic workers are offset when labour productivity and turnover rates are taken into consideration. Caribbean workers were found to be 9 percent more productive than domestic workers. The high turnover rate among domestic workers resulted in crop deterioration of at least 1 percent of the total crop value, but Caribbean workers, on the other hand, tended to stay on the job for the duration of their contract (1991, 113). The same argument can be applied to Mexican workers. Leamington greenhouse vegetable growers made the following comments about the productivity of Mexican workers and the quality of their work: "We know, if we separate crews, if we put four Mexicans on something, it will take a minimum of eight Canadians to get the same amount of work done in the same amount of time. At least double. And then the quality of the workmanship is not there with Canadians, because they don't take no pride in the work. They are not all like that, the majority." And another grower commented: "I think just about any grower will agree with me on ... how good the offshore worker is. It takes 1.5 Canadian workers to do what one offshore does, and that's forgetting that the offshore would work more hours, you know." A similar observation was that "Generally speaking, we've had the odd bad ones, and there is a bad one in all bunches, but in terms of percentages there are as many good offshore workers as there are bad Canadians that would work in a greenhouse. If you have 5 percent bad offshore, you probably have 5 percent good Canadians."

Leamington growers' opinion of local workers is generally low. One grower tells this story about a local worker:

> And I've had three people who said they used to work in agriculture in their country. So I showed them how to take leaves off. I thought they understood 'cause they spoke pretty good English. And when I came back, oh my God, they took leaves off the first five or six plants and then they started taking off tomatoes. And they worked three rows. There were three guys. I didn't know what to do. I didn't know what to say. And I told them give me back the gloves, and I gave them a cheque. I paid them till noon, even though it was ten o'clock, and I said, "Go home." It was probably five hundred dollars worth of damage, if not more. I thought they knew what tomatoes and leaves looked like. And they said, "But we understood that we had to take the tomatoes off." That's the worst experience I've had.

Another grower said he "had two other [local] guys ... working here ... They go to the toilet ... together. There is only one pot and the two guys go together. What's going [on] there? My toilet wasn't working anymore. I go to find out. They had a bottle right in the tank of the toilet."

Productivity among Mexican workers is enhanced through competition among the workers for their employers' favours. Some Mexican workers are co-opted by their *patrones* to ensure compliance with the work discipline among their coworkers. On many farms it is not uncommon for one Mexican worker to be co-opted by the *patrón*. He has often been with the *patrón* for an extended period and can usually understand some English, at the very least at a rudimentary level. The *patrón* often trusts this person to drive a business van. However, this privileged position comes with obligations: in exchange this individual is expected to keep an eye on other workers and report to the grower on any workers who are not meeting expectations. The grower also consults the co-opted worker at the end of the season about who should be named for the following season, but he may even tell the *patrón* to send a bad worker back before the season is over.

Since the number of positions approved by the *Secretaría* is limited, every worker realizes that his chances of getting named by the employer are connected to the chances of other workers. If his *patrón* likes other workers better, he may not be named for next year. Workers therefore often try to outperform each other. Some work too fast in order to impress their boss. Others may try to keep up so that they do not look bad in his eyes, although it is not always possible for them to do so.

But what is most valuable about Mexican labour is that most workers stay until the end of the contract, and while they are in Canada they work long hours, in any weather, every day of the week, even when they get sick or are injured. Ontario growers cannot expect the same degree of commitment to the job from local workers, even from those who are interested in farm jobs, as was discussed in chapter 3. Leamington growers appreciate the fact that offshore workers are available on demand:

> They are here seven days a week, and they work under the conditions, particularly in the summer time, that are in the greenhouse. And they generally don't complain about the work and they are very good workers. They get into a routine. They know their routine, and you can pretty much be sure that the job will get done when you tell them to do the job and you don't have to supervise them hourly.

They [Mexican workers] come in the morning and leave at night and they are here next day. You have to harvest on Sunday and local people won't come on Sunday. You can tell them to come, but you know. You can get away without harvesting on Sunday, but once you get in the summer time ... They are more dependable because they are here. You know what I mean? You won't find anybody in the summer time, and that's our busiest time. You get a lot of excuses from other people but they are just excuses. You know they are not real and you don't get them from those guys. They don't have doctor's appointments (ha-ha-ha). They are here every day. I can't remember the last guy that didn't show up for work, unless he was sick of course.

They don't have to take Johnny to a baseball game or soccer practice ... The offshore workers come here to work. They are happy when they get extra hours.

With respect to the willingness of Mexican workers to work even when it is extremely hot in the greenhouses, the following comments were made:

Usually inside the greenhouses it's 90–95 degrees, 85–90 percent humidity and the guys they come out of there, they are sweating. These guys [the offshore] don't mind. We tell them at one o'clock 'cause it's hottest between one and three. "Take a 3-hour lunch." "No-no, no."

Like the hot weather we had last week, I told my workers, like the offshore, "Why don't you take off and come back in the afternoon, say six o'clock." "Oh no, not a problem. We are OK."

In the opinion of every farmer, offshore workers want to work as many hours as the growers can offer them. The following comments are typical:

When you get a domestic worker, the first thing they ask is how much you're paying. When you get an offshore worker, the first thing he asks is how many hours you gonna ask him to work. That's the difference. You take it from there.

That's why so many people are happy with the offshore workers, because they're living on your farm. You provide accommodations. They are there every morning, and they are happy to have that job.

And the biggest attraction, I think, with the offshore workers is that they are always there. They don't have to take their family to the doctor ...

> but the offshore workers work seven days, wish they work seven days a week. Their attitude is they want as many hours as possible. So it's very easy to set the schedule, especially in a farm-related business, where one week there is not as much work as the next week. Whereas in a factory, a person is working on a shift and the line moves the same speed.

> When I hire offshore workers they come here on a contract ... With those guys if you tell them, "Look we got to work ten hours a day," they won't say no, because they need money. They come here to make money and they need the money and they would work.

> They are good people. They're reliable. First of all, you got them on your farm seven days a week ... There is no problem with them working six and a half days, six days. Like right now, they're putting in good hours. I have no problem with them. They are putting in 127, 128 hours in a two-week period.That's the hours they are putting in.

> These guys come to work, and if they can't work on Sunday they are disappointed. They'll work on average seventy hours a week, offshore. And local people, it wouldn't matter to me if they only worked forty-five. That would be fine. But because [the Mexicans] want to work so bad, then we accommodate them and if they can't work on Sundays they are down in the dumps. No holidays taken ... They are here to make money. They are here to work, and it's worked out very well. Without the offshore labour we wouldn't be the size we are now.

Since growers believe that foreign workers wish to work as many hours as possible, it is not uncommon for them to arrange for migrants to work twelve hours or more each day, seven days a week, during the peak season. Among the workers we interviewed in San Cristóbal, many said they had worked twelve to fourteen hours, six days a week, and half a day on Sunday. Preibisch interviewed a seasonal migrant who claimed that migrants worked up to nineteen hours a day during the harvest season, and one mentioned that they were allowed to sleep only four hours. The latter individual also mentioned that when they were falling off their feet for lack of sleep, they were reprimanded by their employer (Preibisch 1998, 5). As one tobacco worker cited in Preibisch's study commented, they were lucky that the harvest season lasted only two months. Similarly, Smart (1998, 150) found that for three weeks in August in Alberta, Mexican seasonal workers were employed for fourteen to sixteen hours a day, with one or two hours for meal breaks and that fifteen-minute coffee breaks were often withheld.

The agreement for the employment in Canada of seasonal agricultural workers from Mexico stipulates that "for each six consecutive days of work, the worker will be entitled to one day of rest." However, it also states that "where the urgency to finish farm work cannot be delayed, the employer may request the worker's consent to postpone that day until a mutually agreeable date." The latter stipulation allows the growers to expect Mexican farm workers to work seven days a week, including half a day on Sunday, without any "mutually agreeable" day of rest being offered to them during the peak season. One tobacco worker cited in Colby's study (1997, 16) complained that "We worked right through, and then the day came for us to go to the airport, and the days were lost. We got paid well, but we didn't get the rest." Even when they get sick or are injured, Mexican workers do not want to ask for days of rest. Their unwillingness to lose income is only one reason why they return to work.

COMPLIANCE

The most important reason why Mexican workers are always available for work is their fear of being expelled from the program by the Mexican Ministry of Labour. Farmers are requested to write a report on each worker at the end of the season. Placed in a sealed envelope, it is to be handed in by workers to a *licenciado* at the *Secretaría* upon the their return to Mexico. If the farmer reports that the worker in question is lazy, does not get along with others, drinks too much, or is slow or rebellious, he or she will have a hard time getting approved for the program in the following season. If a *patrón* wants the workers to put in extra hours on Sunday (the only day when they get at least half a day off to relax, go to church, do their laundry, and go shopping) and they refuse, they may not be invited to come next year. If a worker choses to leave Canada before the contract is completed (be it for health reasons or for family-related reasons), he or she risks being *castigado/a* (penalized) by the Ministry of Labour the following season. The same happens if a worker approved for work in Canada wishes to take a break from the program for one season or asks for a transfer to another farm. The following season the worker is likely to be *castigado/a.* The punishment does not usually last more than one season, although in some cases it may involve a longer "sentence."

The fear of not having their contracts renewed forces many Mexican workers to return to work even when they are sick. Colby (1997, 16) cites one worker as saying, "My side hurt for weeks. I thought I broke some ribs. My boss told me to take a day in bed. So I worked again, even though I hurt. I didn't want to cause problems because I need to come back next year. I have five children at home." Preibisch (1998, 7) also cites one worker who complained, "When one starts getting sick and goes to see a doctor frequently, the *patrón* does not ask for us any more. That time my lower back was in pain and I was sick for a while but I didn't complain to the *patrón*. For if I tell him, he does not ask for me any more and I am left without work. That is why I put up with it until I returned here to treat it" (the translation is mine). Some Mexican men interviewed in Leamington have had similar experiences.

Rufino, who was employed on a farm in Essex County, had high blood pressure. He brought some pills from Mexico but ran out of them quickly. The stress he experienced at work worsened his condition. Yet he did not dare to ask his *patrón* for a day off to see a physician. He kept on working until he returned to Mexico.

Porfirio, on the other hand, could not wait to complete his contract:

Porfirio got sick as soon as he came to work in Ontario. He continuously had headaches. He went to see a doctor and was told that there was nothing wrong with him. Yet he had high blood pressure and felt anxious. The physician told him to take Tylenol. Porfirio continued to feel sick. Without a letter from a physician confirming his state, Porfirio decided to use his wife's medical condition as an excuse to return home. He was released from work, but the following year he was penalized. Only in the year after that was he allowed to participate in the program again.

Mexican workers return to work even after having an accident, provided they are still capable of performing the job. At times, they continue working in spite of the pain or discomfort they experience, or they take a few unpaid days off. Barrón (1999, 121) cites one Mexican female worker employed in the Niagara region who said, "I fell through a glass and had to have my leg stitched up, but when they told me I had to rest for two days, I went back to work so I wouldn't lose the money." Even though the workers are entitled to days off paid for by the Workplace Safety and Insurance Board, many of them are ignorant of their rights or do not wish to look bad in the eyes of their *patrones* by claiming the benefits of the insurance. The story of Don Prudencio, a Leamington-area interviewee, provides an illustration:

Don Prudencio, a fifty-year-old Mexican worker, was sprayed with a pesticide while he was working in a greenhouse. Not having been instructed on the first aid measures he was supposed to take, he continued working. On the second day, he felt an unbearable burning in his eye and reported it to the supervisor. It was only late at night that the supervisor had the time to take Don Prudencio to the hospital. By then much of his cornea had been destroyed. He was prescribed eye drops and went back to working in the greenhouse. Gradually he started losing vision in one eye. Then his other eye started burning and tearing. Even then he continued to work.

The unofficial practice of loaning workers to other employers aggravates the problem. Canadian farmers are in fact prohibited from doing so without written consent from the HRDC and the government agent, but in practice, growers who receive requests for supplementary labour from other farmers ignore this bureaucratic procedure.* One consequence is that workers who experience accidents when working without permission have no recourse. The story of Adrian, another Leamington-area interviewee, provides a good illustration:

Adrian, who was authorized to work in a greenhouse, was "loaned" by his *patrón* to his cousin, who needed extra workers to help him pick apples when the harvest in the greenhouse was over and there was very little other work left to be done in it. While picking apples, Adrian fell off a ladder and broke a rib. He was in pain for weeks, yet he did not complain either to his own *patrón* or to his cousin. Nor did he take any time off. He continued working *aguantando el dolor* (while putting up with the pain). And it was only when he returned to Mexico that he allowed himself the "luxury" of seeing a doctor, who informed him that he had a broken rib.

*The unauthorized loaning and trading of workers is also reported by Colby (1997, 16).

Another reason Mexican workers report to work every day if they are needed and do not ask for time off unless they are seriously ill is that they do not have a social life outside work. They leave their families behind. They have no birthday parties, funerals, or weddings to attend. They do not need to take their children to a game or to a doctor. They come to Canada to work and not to have fun. And even when they do go out to a pub on a Friday or Saturday night, they try to get home early so that the next day they are in good shape to work as many hours as are required of them. There is very little social life for Mexicans in Canada. Occasionally they visit each other, and at times they hang out at the entrance to a local supermarket. In Leamington, Mexicans gather at the entrance to the No Frills grocery store on Thursday or Friday nights and on Sunday afternoons. In other communities other supermarkets are patronized by Mexican workers. In some communities there are Sunday services in Spanish, and in some communities Mexican workers celebrate the Mexican Day of Independence with a dance or a procession and with traditional *gritos* (screaming *Viva Mexico*! and other "*vivas*"). But in some communities Mexicans have no social life at all. As one Leamington grower explainded, "The offshore just come here to work. There are no other commitments. They come here to make money and to take it to their family." Similarly, another grower observed that "They do the job without complaints 'cause they come to this country for one thing and one thing only – to make money, to go back home, to live good, and to come back here next year to make more money."

This level of commitment to the job cannot be expected of domestic workers, even the most reliable of them. Domestic workers also have church and family commitments. One grower reflected on, "that aspect of domestic workers, there are family things, vacations, [important] times [for] the Mennonite population ... like last Sunday there was a baptism, so no one comes to work, last week there was something else. The bulk of our domestic people are Mennonites."

In some way, the *patrones* discourage their workers from having social activities outside of work. A worker who returns home too late or who gets late visitors, particularly women, risks earning the disapproval of his *patrón*, who believes that such activities might exhaust the worker and make it difficult for him to work well the next morning. Cecil and Ebanks (1991, 399) report that the overwhelming majority of the farmers employing Caribbean offshore workers interviewed in their study disapproved of dating and friendship between West Indian men and Canadian residents. In the growers' eyes, those who wished to spend money on having fun were not good workers; the good workers were those who came to Canada strictly to make money in order to improve their living standards at home (Larkin 1990, 55). I suspect that growers hiring Mexican workers feel similarly about their offshore workers' social life.

Finally, paternalistic relationships between Mexican workers and their *patrones* cement the workers' loyalty to their employers and commitment to their jobs. As many researchers have pointed out, paternalism is an important mechanism through which employers ensure their workers' loyalty in situations of labour shortages. Paternalistic practices may include provision of housing, credit, and subsidized transportation to and from work, sponsorship of festivities, extension of personal favours and rewards, coverage of medical expenses, and visits of migrant farmers in their home communities (Wall 1992; Stultz 1987; Pentland 1981; Alston and Ferrie 1993; Laliberte and Satzewich 1999). As Terry contends, while such practices may be well-intentioned, they reinforce "the feudal serf-master relationship that colors the farmer's dealing with his migrant workers" (1988, 78–9).

There is, indeed, a feudalistic, paternalistic relationship between the *patrones* and their Mexican workers. Workers reside on the premises, often next to the houses of their *patrones*. The *patrones* or their wives take the Mexican workers shopping every Thursday or Friday, or they give them money to take a taxi into town and back. When Mexican workers get sick, it is, again, the *patrones* or their wives who ensure that they receive medical attention, whether it involves a doctor's visit or simply purchasing medicine. Any medical costs incurred are deducted from the workers' pay cheques. Use of the telephone in the *patrones* home; coffee, soft drinks, or fruit offered during breaks; and assistance with money transfers at the bank are all ways in which the farm owners assist their workers. Some workers are taken on short trips in Ontario or to other sites in Canada with their *patrones*. Occasionally, farmers will learn basic Spanish, particularly on smaller farms where farmers work side by side with the Mexican workers. A knowledge of Spanish allows them to chat and joke with their favourite workers.

Some Canadian farmers (or other members of their households) even visit their workers in Mexico.

Because of the isolation of the work environment and housing arrangements, Mexican workers are excluded from the social world of the Leamington community. As such, they come to see their *patrones* and their *patrones*' families as offering the only opportunity for meaningful social interaction within Canadian society, apart from the impersonal verbal exchanges at banks, stores, or fast-food outlets. The Mexican workers' interaction with local residents is not always pleasant. Marcelino commented that "some people ... treat us well, but ... others look down on us." Leonardo agreed: "There are many people who look down on us, don't think we are worth anything. Some of them don't want us to work here." And Ignacio added that "People in Leamington think all Mexicans are drunks." Don Vicente, who has been coming to Leamington for fifteen years, elaborated:

> Even though they used to be nicer before, people in Leamington never really liked the Mexicans. They think we are all thieves. Perhaps, some of us are. And in Mexico even our president is a thief. But you know how it is in small towns. People trust each other. And in *ranchos* even more so. Here we go to Zeller's [and Canadian Tire – adds his room-mate] and they stare at us to make sure we won't steal. The same when we go to yard sales. Because I speak some English, I chat with them sometimes. But they all stare at others with fear. There are some pubs where we are not allowed. Because I like dancing, I used to go to pubs when I first came over, but women don't even want to dance with you. And they are rude. I understand it when a guy is drunk, and he does not take no for an answer. But when you approach a girl and ask her to dance with you politely, there is no reason for her to be rude. One woman even punched a Mexican guy in his nose when he asked her to dance with him. His nose bled.

By comparison with the local residents involved in these experiences, the *patrón* is a friend who does not look down upon his workers and who can assist them in times of trouble. And so Mexican workers try to please their employer because of this personal commitment to him and his family, and they try to be available for work whenever the need arises.

THE SIGNIFICANCE OF THE OFFSHORE PROGRAM FOR LEAMINGTON GROWERS

Thus, to reiterate, through the Seasonal Agricultural Workers Program, Ontario growers get the captive labour force that they require. The workers who are sent to their farms are not free to leave their jobs. They are available for work in the evenings, on holidays, and on weekends, and they rarely take time off, even when they are sick or injured. In other words, they are always there when the growers and their crops need them. The growers clearly see that the availability of such labour allows them to survive and even expand, as the following comments made during the interviews for this study illustrate:

I wouldn't even be here if I didn't have offshore labour. I would sell ... at a loss and ... do something else. Everybody's operation is dependent on them. It's come to the point where it's totally impossible to do without this offshore labour. I don't know anybody that does.

We would be in deep trouble if we didn't have the offshore at this point, unless someone comes up with something. Deep trouble, deep trouble.

Well, if it wasn't for the offshore worker, the greenhouse industry would have a lot of unemployed people. That would solve the problem, maybe.

If we didn't have the offshore labour, the expansion wouldn't be there.

If you take the offshore program away from this area, you can effectively shut down the greenhouse industry, 'cause you won't get the labour force.

If it wasn't for our offshore program, honestly, we would not be able to survive.

We could not operate without them, no chance.

And the growers realize, of course, that what makes Mexican (and Caribbean) labour so valuable is that the workers are not free to quit. If Mexican workers were to come as permanent residents, this advantage would be lost, as one grower points out: "The disadvantage of legalizing Mexicans would be that they would be free to leave the greenhouses and go to work cutting mushrooms, for instance. Mushrooms are 'high tech' now. They are air conditioned – sixty degrees. You'll get workers there. They will be there before they'll be here.

SUMMARY

Offshore workers recruited through a FARMS-administered bilateral program are not necessarily cheap. While they agree to work for a wage that is only five cents above the minimum hourly wage, other expenses associated with the program such as the transportation costs and the provision of housing make them more expensive than local workers. But these extra expenses are well compensated for by the workers' high productivity and compliance with their working conditions and, most importantly, by their loyalty. They usually stay until the end of the season and are available for work every day of the week. They rarely take days off even when they are sick or have suffered accidents.

Their loyalty is ensured mostly through their fear of being expelled from the program, since their *patrones* have the power to decide not only whether they will come back to the same farm the next year but also whether they will even stay in the program. The *patrones* are required to send reports to the *Secretaría* to help the Ministry of Labour and Social Planning to select participants, and the workers are "punished" if they are reported to lack commitment to work by their Canadian employers.

Additionally, lack of social commitments outside work makes these migrants, separated from their families, their kin, and their communities, available for work whenever there is a demand, as does the paternalistic relationship they develop with their *patrones*, who are often perceived as "good guys."

One might expect that after having worked in Canada for one or more seasons Mexican workers would be in a better economic position in their home communities and less desperate to agree to work for minimum wages. But they are not. As I argue in chapter 8, the potential for economic growth related to productive investment in Mexico is very limited. First, most Mexican seasonal workers spend their remittances on consumption (building a house, buying appliances, and clothing their families), the education of their children, and medical services. Second, while some do purchase land or small businesses, these ventures offer no more than supplementary income and depend on continued financial support from remittances earned in Canada. Small plots of land of no more than three hectares and usually lacking irrigation allow Mexican workers who buy them to grow some subsistence crops such as corn, but without adequate support from the banks and the government, the further productive growth of these small farms is extremely limited. Thus, while the standards of living of Mexican seasonal workers and their households do in fact improve, the maintenance of this life-style depends on sustained earnings in Canada. For fear of losing their chance to return to Canada Mexican workers comply with their employers' expectations and are available for work whenever they are needed.

- LABOUR IS RELIABLE & DOCILE
- MEXICANS GET LESS THAN THE PREVAILING FARM WORK WAGES, AGAINST STIPULATIONS
- MX CONSULATE REPS. EITHER DO NOTHING TO HELP WORKERS OR TAKE THE SIDE OF FARMERS, AS IF THEY WERE A WHITE OR COMPANY UNION. SOME FARMERS PAID BRIBES SO THAT CONSULAR OFFICIALS WOULD NOT REPORT THEIR VIOLATIONS TO CDN. AUTHORITIES.
- MX WKRS ARE MUCH MORE PRODUCTIVE THAN CDN WKS. "AT LEAST DOUBLE," SAYS ONE GROWER, OR "IT TAKES 1.5 CDN TO DO THE WORK OF ONE MEXICAN, SAYS ANOTHER. OR "IF YOU HAVE 5 PERCENT BAD OFFSHORE, YOU PROBABLY HAVE 5 PERCENT GOOD CDNS." (116 FOR ALL QUOTES).
- MX WKRS WILL STAY THROUGH END OF SEASON, WK. LONG HRS, IN ANY WEATHER, EVEN WHEN SICK OR INJURED. (117).

(CITE P. 118)

- CDN. RACISM MAKES CDN FARMERS LOOK GOOD BY COMPARISON.
- (CITE P. 126) - INCOME IS JUST FOR REPRODUCTION, I.E. NO PRODUCTIVE INVESTMENTS.

8 The Migrant Syndrome

At long last Rodolfo was on his way home. He missed his wife, his three children, and his parents. Rodolfo had spoken with them by phone once a month, usually after he had sent money to them *banco-a-banco*. Even though it had cost him more than twenty dollars to send the money, it was a more secure method than sending a money order by mail. If the letter was lost, what could you do? Not that bank transfers always worked well. His friend Marcelino had spent a lot of sleepless nights and much money on phone calls to Mexico because the bank had sent his money to the wrong address. Rodolfo had usually sent one thousand Canadian dollars to his family once a month. His wife, Celia, had been very nervous the first time she had had to go into town to receive a bank transfer. She had brought her brother along to help her. But then she had learned to do it by herself. Sometimes Celia had cried on the phone. She had told him that she missed him and that it had been difficult to manage the children without their father around. But at the same time she had been glad to receive the money that she needed so badly to manage the household.

Looking back at the eight months that he had spent in Canada, Rodolfo thought that they were certainly tough. Yet it felt good to know that he could provide for his family, that his eldest son would be able to attend a *secundaria*, and that he would be able to start buying material to build a house. He would not start building a house until next year (if he was sent to Canada again). But at least he could buy *tabique*. And he could pay off the debts for his wife's caesarian section. And he had brought nice gifts for Celia and the children. He had even bought a television and a CD player. Next year, if everything went well, he would buy a VCR. And after he finished building his house, maybe he would buy some animals – a cow, some pigs, chickens. Maybe, if he was lucky, he would even be able to buy a hectare of land to grow a bit of corn, so that Celia could make her own corn flour for tortillas. But it was better not to think that far ahead. *Dios primero*, if he stayed healthy and his *patrón* asked for him, he would be able to return to Canada next year. Rodolfo made the sign of the cross and prayed for his journey home to be safe. And then he fell asleep. Tomorrow he would be at home.

As I argued in the previous chapter, it is possible to hire Mexican seasonal workers cheaply because of their fear of being *castigados* and thrown out of the program. Participation in the program is highly valued, because for most Mexican workers there are virtually no alternative sources of income that would allow them to earn nearly as much. As I have suggested in chapter 6, most program participants are landless *jornaleros* when they first apply to join the program. But despite the relatively high levels of income earned in Canada, for most of them their occupational status remains the same even after they have worked in Canada for several years. Money earned in Canada is used to build houses, to pay for medical services and school, and to purchase cars, clothes, and appliances. Very little of this income is invested in productive activities that could cover the household's expenses in the future. Once the family has decided to send their children to school, it is essential for the male heads of the households to renew their Canadian seasonal work visas for many more years – up to thirty years – until all their children have received the education they require. At the same time, once the household's standards of living improve, its members' expectations rise. Yet, without continued seasonal migration to Canada, it would be impossible to meet these expectations. Participants in the Canadian program become locked into a cycle of transmigration that Reichert (1981) has coined the "migrant syndrome." Because of the need to reproduce this transmigratory pattern, Mexicans are willing to work as many hours as are required.

As mentioned, very few Mexican workers have invested their Canadian earnings in land or businesses. Ironically, those who have made productive investments are people from urban and semi-urban areas, people who would not have been selected to participate in the program if the criteria set by Ministry of Labour had been rigidly applied. On the other hand, a typical participant – poor, uneducated, and originating from a *rancho* with limited employment opportunities and scarce land lacking irrigation – has very dim prospects for investment. This chapter will illustrate how Canadian dollars are spent in Mexico and attempt to explain the patterns. As mentioned earlier, I conducted research on the use of remittances by Mexican seasonal workers in several stages. First, 154 Mexican workers were interviewed in the Leamington area. Among them were some who originated in small towns and semi-urban areas around Mexico City, despite the Ministry of Labour and Social Planning's stated preference for participants with rural backgrounds. Then 100 participants in the Canadian "offshore" program were interviewed in the village of San Cristóbal. The uniqueness of this village lies in the sheer number of men working in Canada on a seasonal basis. Finally, a survey of 311 workers was conducted in eleven villages in the states of Guanajuato and Tlaxcala. In this chapter I will at times report on the entire sample; at other times I will draw a distinction between the three subsets, and yet at other times I will discuss only some of them, in order to avoid repetition.

USE OF EARNINGS BY CANADIAN MIGRANT WORKERS

Working, on average, fifty hours per week, Mexican farm workers employed in southwestern Ontario earn $Can345 ($U.S.240) per week. If they work in Canada for eight months, they can send home some $Can8,000 ($U.S.5,500), or over 40,000 Mexican pesos. For them this salary is a fortune. As discussed in chapter 6, Mexico has a high rate of underemployment, particularly in rural areas. Many Mexicans work as *jornaleros*, but their jobs are generally seasonal, sporadic, and poorly paid. Even if a *jornalero* is lucky enough to work six days a week for fifty-two weeks, he will earn the equivalent of $Can1,500–1,600. While nonagricultural wages are higher, they are still considerably lower than the income earned in Canada. On average a Mexican worker can send home $Can1,000 or more every month. Thus, in eight months a Mexican migrant can send home approximately $Can 8,000, or an equivalent of five to six years' salary. In the first year, earnings are generally used to maintain the migrants' households and pay off the debts they have to incur in order to join the program. As mentioned, the costs include bus fare to Mexico City, hotels, and medical examinations. It is often not until the second year that migrant workers start investing their remittances. They spend their money mainly on building or repairing a house and on their children's education, while very few use their earnings to buy land or a small business.

Housing

An opportunity to work in Canada has allowed most participants in the Seasonal Agricultural Workers' Program to improve their living conditions. Many were still living with their parents or parents-in-law when they started working in Canada. The first thing they wanted to do after paying off their debts was to build a house where their family could live independently. Some had to buy land for a house first. Others were given a lot on their parents' property. Then they would buy *tabique* (cinder block), hire carpenters, and start building a house little by little. First just one room, then another room, then another storey, then a kitchen, and so on.

Depending on the remittances they brought home, it might take from two to seven years just to have what the family needed. And then it would still be possible to keep adding rooms to the house for years and years. In 1997 a piece of land for a house in the countryside in Guanajuato cost 8,000–10,000 pesos (U.S.$1,000 – 1,250), but in other areas the price could be as high as 30,000 pesos (U.S.$3,750). Among those interviewed in Leamington, 122 respondents (or close to 80 percent) had purchased houses or made improvements to their houses using money earned in Canada. The same was true for the other eleven communities we visited in 1999 and 2000. The village of San Cristóbal was somewhat different; there, every program participant that we interviewed had bought or improved his family's house.

Children's Education

Children's education is also very important to Mexican migrant workers. For many, the fact that they can send their children to school is a source of great pride. Not having had an opportunity to go to school themselves, many of them feel that they owe it to their children to do everything possible to better prepare them to compete for jobs. Some Mexicans interviewed in the study mentioned to me that to get a job in Mexico City, one needed to have a secondary school diploma. There are primary schools in virtually every Mexican village, and while primary school education is free, parents do have to pay for their children's secondary and postsecondary education. Many *ranchos* now have either a secondary school or a *telesecundaria* (distance-education secondary school). Alternatively, there are *secundarias* and *telesecundarias* in neighbouring *ranchos*. In this case, parents are expected to cover the transportation costs required for their children to attend school. In addition to tuition fees, other education-related costs include books, uniforms, and school supplies.

Among those interviewed in Leamington, 79 (or 51 percent) told me that they had used the money earned in Canada to send their children to a secondary school, *preparatoria* (high school), or university. Among the 100 workers interviewed in San Cristóbal, 34 reported having done so. The difference between the Leamington-area respondents and the workers interviewed in San Cristóbal is related to the composition of the former. As mentioned earlier, some workers employed in the Leamington area were from urban areas where more schools were available and where parents might have had higher aspirations for their children's education. However, in both groups the actual investment in children's education may have been underestimated by the male workers answering the question.

At the last stage of this project we used a different method to estimate the degree to which the workers' incomes had been spent on their children's education. Rather than asking them or their wives whether Canadian-earned money had contributed to their children's education (as was done at the earlier stages of the project), we asked the respondents to list all their children and their levels of education. If at least one of the worker's children had received education beyond the primary level since the worker's first trip to Canada, I coded "investment in education" as "yes"; otherwise it was coded as "no." Using this methodology, I calculated that 60 percent of the workers had contributed to their children's education. It is possible that since household budgets are generally managed by women, male migrant workers may be unaware of how the money they send to their wives is used and that they are therefore likely to assume that their remittances are not spent on their children's education, whereas in fact they are.*

Productive Investment

Only 141 of the 565 Mexican farm workers interviewed in the study had invested their earnings in productive activities. The respondents who had not been able to invest their remittances productively complained that agricultural land was too expensive or unavailable in their region. Anselmo, from Guanajuato, commented, "Those who have land do not want to sell it. Instead, they would rather buy more. And those who sell it set very high prices."

Similarly, many feared that it was difficult for small businesses to survive in the current economic climate. Rufino commented that "It is difficult to open a business, because since 1982 the country has been in an economic crisis. Before, yes, one could sell, have a business. But today it won't work. One will invest and earn nothing." Flavio, reflecting on the dangers of investing money in raising livestock, commented, "Sometimes it is possible to invest badly, because if animals get sick, you are finished."

Yet 141 respondents had used their earnings to buy land, livestock, or a small business. Among those interviewed in Leamington, a total of 33 participants had used the money they had earned in Canada to purchase land, livestock or a small business. Fourteen had purchased land, 5 had bought cows, horses, pigs, or chickens, and 17 had invested in such businesses as real estate, a shoe store or shoe sales, a moving truck, an electrical repair shop, a tractor for working their own or other people's land, a taxi, a tire shop, a tailor's shop, a lamp and lighting store, a blacksmith shop (specializing in doors and windows), or the sale of clothing. Some among them had combined several forms of these types of investment. Among the 27 migrant workers interviewed in San Cristóbal who had invested productively, 4 had purchased one or two hectares of mostly rain-fed land lacking irrigation, 10 had purchased livestock (3 of them combining it with the purchase of a truck or tools), and 16 had made business-related purchases of such things as a food store (7 people), tools for workshops (5 people), construction

*For a fuller discussion of the impact of Canada-bound seasonal workers' remittances on their children's education, please read Basok (2000a).

equipment, and trucks (3 people) to be used for transporting either people or fertilizers.

The longer a migrant works in Canada, the more likely he or she is to invest productively. While the average number of years spent in Canada by Mexican workers interviewed in Leamington and in San Cristóbal was 6.5, for those who had invested remittances productively, it was 8.4. Yet, among the 311 workers interviewed at the last stage of the project, the difference between the two groups was not as clearly demarcated (6 and 7 years respectively). Even after having worked in Canada for many years, most program participants could not invest their money in activities that would allow them to retire from the Canadian program. Most considered these investments as supplementary, and they felt they had to continue working in Canada to meet their family needs. Only two seasonal workers from San Cristóbal who had purchased businesses no longer participated in the Canadian program at the time they were interviewed. Among those who invested in small businesses, many worked in them only when they returned from Canada. Such businesses included masonry, blacksmith shops, a tailor's shop, the operation of a tractor, and an electrical repair shop. Most construction work is done in Mexican villages around Christmas time, when migrant workers (both U.S.- and Canada-bound) return home with cash. Thus, for a migrant worker who rents construction equipment, the season of high demand for his business corresponds to the period of his return from Canada. Other seasonal workers who did operate their businesses throughout the year (businesses such as stores, a fruit stall, house rental, moving, a shoe store, and shoe sales) relied on the labour of their wives and grown-up children, mostly sons, to run them. One migrant worker had no intention of working in the businesses he had purchased: he had bought a taxi for one of his sons and a tire shop for another. Of course, many migrant workers who have grown-up children available to help them prefer them to go to school and obtain an education that would improve their chances of experiencing the socioeconomic mobility that small businesses and agricultural investments are unable to offer. Of 18 people who had made agricultural investments, 9 relied on the assistance of their sons (or father, in one case). Two others who had owned land before coming to Canada and who had invested in agricultural implements hired *jornaleros*. The option of hiring *jornaleros* is not available to many others, because small-scale subsistence production in a rain-fed region is vulnerable to price and weather fluctuations. If the crop is lost or sold cheaply, no money is left to pay *jornaleros* for their work. Neither land nor livestock nor small businesses purchased with money earned in Canada gave their owners enough confidence to give up their trips to Canada. They often permitted their family members to earn some extra cash or provided them with an occupation upon their return to Mexico, but the major source of income was still Canadian agriculture.

OBSTACLES TO PRODUCTIVE INVESTMENT

Mexican migrants to Canada resemble other seasonal migrants in their investment-related decisions (see Reichert and Massey 1982, 8–11, for a summary of cross-cultural research on remittances). Reichert and Massey point out that migrants from a wide variety of countries invest little or none of their incomes in capital-generating activities in their home communities and that capital investments made by migrants typically constitute secondary or tertiary economic activities that merely supplement migrant labour, without replacing it (9). They observe that even when migrants purchase land, they rarely view it as a means of achieving local self-sufficiency that could obviate the need for continued migration. Instead, land is purchased as a speculative investment to supplement foreign income. The purchases add to the prestige of the owner, since in these communities possession of land is the traditional expression of security and wealth (9–10).

Research on U.S.-bound Mexican migration points out that migrants originating from urban and semiurban communities are more likely to invest productively than their rural compatriots. Furthermore, migrants from rural areas with good soil and infrastructure or that are undergoing development are more likely to invest their earnings in productive activities than those originating from less-developed rural communities (Alejandre et al. 1991, 73–5; Cornelius 1991). Finally, a migrant who already has economic resources (land or a small business) that can be expanded and reinforced with the help of "migradollars" is more likely to use U.S.-earned money in this way (Massey and Basem 1992). The irony of the Canadian program, however, is that few participants originate from such communities or have such characteristics. The objective of the Ministry of Labour and Social Planning is to assist those who need it the most, and therefore most participants are not business owners or landowners, and they come from areas with a shortage of arable land for sale and a poor potential for investment in small businesses. Additionally, most have low levels of education. Without education they may find it difficult to manage a small business.

Canada-bound migrants have a certain edge over their U.S.-bound compatriots. Their jobs are secure, and they do not have to pay rent. Thus they are more likely to remit more money back to Mexico. In addition, as we have seen, because they are sent to rural areas devoid of Latino cultural activities and because they work long hours up to seven days a week, they end up spending very little of their earnings and remitting most of them. At the same time, most do not invest their money productively because they originate from communities with little potential for investment. Like U.S.-bound migrants, the migrants to Canada who do manage to invest productively are often from urban and semiurban areas, or they have nonagricultural skills and some resources. Among the seventeen respondents from the Leamington sample who invested their remittances in small businesses, only two tractor owners, one tailor, and one owner of a lamps and lights store (which, incidentally, went out of business) were from rural areas. The rest were from urban and semiurban areas and predominantly from Mexico City. Of course, if the Ministry of Labour had used the established criteria rigidly, perhaps these people would not have been selected to participate in the program.

Mexican workers seem to be dependent on the renewal of their contracts more than Caribbean participants in the program. When asked to explain reasons for participating in the program, virtually all interviewed Mexicans mentioned being poor and not having a steady job in Mexico and, as a result, being unable to build a house or send their children to school. By contrast, for West Indian men, seasonal work in Canada, or "workin' on the contract," as they call it, has a rather different meaning. Larkin (1990, 28) lists various reasons why West Indian men participate in the Agricultural Seasonal Workers Program. Gaining experience is an important consideration for them. Some of them call it a working holiday, while others refer to it as a rite of passage. They appreciate the male solidarity in the bunkhouse and the opportunity to learn new farming techniques and different social customs. They also value their interaction with white society and hope to change relations between blacks and whites.

The knowledge and experiences these men talk about can be gained in a few years, and after that they no longer need to work in Canada, though in practice many of them end up returning (Cecil and Ebanks 1992). Compared to Mexican participants in the program, they are not as poor. They are relatively well educated, and as many as 40 percent of the Caribbean workers interviewed by Cecil and Ebanks claimed ownership of farms, although the term actually used in the interviews was "ownership of land," which might have referred to various land tenure arrangements (23–4).

In sum, the very criteria used by the Mexican Ministry of Labour to select participants for the Canadian Seasonal Agricultural Workers Program inadvertently ensure that these workers do not become landowners or successful business owners and instead, by growing dependent on incomes earned in Canada, become locked into the cycle of transmigration. Since it is important for most Mexican workers participating in the Canadian program to return to Canada in the years to come, they do not wish to jeopardize their continued participation by refusing to work when their *patrones* demand their labour.

9 Summary and Conclusion

Mexican seasonal farm workers hired through a government-regulated program play a crucial role in Canadian agriculture. They are brought in to fill the demand for labour in agricultural sectors that have experienced significant problems trying to recruit and retain domestic workers. Several trends have contributed to a rising demand for labour in Canadian agriculture since World War II. Among them are the consolidation of farms, a decrease in the size of growers' households, and the declining interest among the children of growers in farming as an occupation. Canadian growers have experienced serious problems meeting this rising demand from the pool of domestic workers. Various labour recruitment schemes initiated by public and private agencies have delivered workers to the farmers, but these people have rarely stayed on the job long enough for the crops to be harvested or for the farmers to feel that the time they have spent training these workers has been well invested. The only policy initiative that seems to have satisfied the needs of the growers is the Commonwealth Caribbean and Mexican Seasonal Agricultural Workers Program. Some growers feel that the offshore program, as it is commonly known, is so vital to the industry that without it many farms would have folded. In essence, as I will explain more fully below, Caribbean and Mexican offshore workers have become a "structural necessity" for Canadian fruit, vegetable, and tobacco growers.

The major reason why the survival of the industry hinges on the recruitment of these workers is that they are hired as unfree labour. Not only are they unfree to circulate in the labour market, but they feel strongly compelled to stay until the end of their contracts and to be available to work on demand. By and large, growers can expect them to work long hours, seven days a week in any weather, even when sick or injured, until their services are no longer needed. The perishability of the crops to be harvested makes it imperative for the growers to have workers available when the need arises. Because of the trends mentioned above, farm operators and members of their households can no longer furnish all the required labour when emergencies arise. At the same time, it is extremely difficult to find workers who would be as committed to the survival of the crops as the farm operators themselves and who would therefore be willing to forego family obligations, religious commitments, and social responsibilities in order to provide the required labour on demand. The only workers who would be in a position to meet the growers' demands for readily available labour would be those who felt compelled, for whatever reason, to please the employer and who had no social commitments outside of work. Mexican seasonal workers hired through the offshore program meet these criteria. Mexican workers are unfree, not only because their contracts tie them to one particular employer and deprive them of the right to change jobs but also because they feel obligated to provide labour whenever their services are required.

Many Mexican participants in the offshore program are recruited among landless peasants from small rural villages with no alternative sources of income that could sufficiently meet numerous needs of the migrants' households. An opportunity to work in Canada and earn in one season an equivalent of a salary earned in Mexico in five to six years is highly valued by many Mexicans, who are willing to part with their families for up to eight months each year. Most of the income earned in Canada is sent or brought home to cover vast household needs: children are sent to school, medical care is provided to household members, and the family is better dressed and fed. A single trip to Canada cannot meet all these needs. It takes many years for children to get meaningful education (a secondary school diploma, at least). Furthermore, as the household members' standards of living rise, so do their expectations. This makes it imperative for the migrant to continue working in

Canada. And even those who buy small plots of land or businesses feel that their investments are neither secure nor significant enough to provide their families with sufficient income to meet their rising expectations. While illegal migration to the United States is an alternative available to Mexicans interested in earning a higher income, most participants in the Canadian program do not consider it acceptable. The high costs of travel to that country and the fear of being apprehended by the *migra* (immigration officers) discourage many from going there. By contrast, travel to Canada is financed by the employers, and temporary residence is authorized by the Canadian immigration department. In addition, many Mexican migrants prefer the Canadian program because it offers secure employment, free and stable housing, and fewer temptations to spend the earned income frivolously.

Whether Mexican workers are allowed to continue participating in the offshore program depends to a large extent on the evaluation of their performance by the growers to whom they are assigned. Growers prefer workers who are productive, compliant, disciplined, punctual, and available for work on demand. Workers who do not meet these expectations are not likely to receive good evaluations, much less to be "named" to work on the same farms in the following season. Consequently the fear of receiving a negative evaluation compels Mexican workers who are interested in returning to Canada, if not to the same farmers, to do their best to win a good reputation among the employers.

Winning a good reputation means being available for work on demand. In part, Mexican migrants prefer to work many hours to earn as much money as possible while in Canada, so that they can cover the extensive needs of their households. On the other hand, they do become exhausted when they do not rest adequately. As one Mexican worker interviewed in the study joked, "Mexicans complain all the time. When they work little they complain, and when they work long hours they complain." By and large the sentiments of Mexican workers towards working long hours seven days a week vary throughout the season. When they first arrive in Canada, usually several months before the harvest begins, when there is no need for them to put in long hours, Mexican migrants wish they could work more. By the end of the harvest season, however, they wish they could work less. While farmers assume that Mexicans are happy to work all the time and rarely ask for their workers' opinions, I have talked to many Mexicans who wished they had been allowed to rest more often. Some of them, who were employed in a very large greenhouse, even refused to work one Sunday afternoon, despite their fear that their refusal could have negative repercussions. Such incidents of resistance are generally rare, however. No matter how much Mexican workers resent the situation, most return to work when they are asked to, for fear of displeasing their employers and consequently getting poor evaluations.

In order to get positive letters of evaluation, many workers also remain silent in the face of abuse. As we have seen, some Mexican workers are exposed to poisoning because growers violate regulations with respect to pesticide application. Some growers send their workers to spray without offering them protective clothing; others expose them by spraying the plants while they are working right next to them. Yet Mexican workers rarely object to these practices.

They also rarely take time off when injured on the job. Although Mexican workers are entitled to receive workers' compensation for job-related accidents, very few of them do. Most continue working despite the pain or discomfort they experience. When they do take a few days off they lose pay, and for several reasons most workers do not collect insurance benefits, because, first, most are not aware of their right to collect them or of the procedures involved. Neither the Mexican consulate nor the growers provide the necessary information to them. Although physicians in Ontario are required by law to report work-related accidents to the Workplace Safety and Insurance Board (WSIB, formerly known as the Workers' Compensation Board) language barriers prevent them from understanding that a worker's ailment is work-related. Even when physicians do send the required forms to the WSIB, the workers may not make a claim, because by the time the forms arrive at the farmer's address, the workers may already be in Mexico.

The growers make no effort to forward the forms to their workers. In some cases, growers even send injured workers back to Mexico, claiming that they have run out of work and not that their workers have been incapacitated as a result of an accident or pesticide poisoning. This makes it impossible for the injured workers to claim compensation for lost time. What makes the situation even worse is that in Mexico the returnees have to pay for private medical services out of their own pockets, whereas if they had stayed in

Canada they would have been covered by the Ontario Health Insurance Plan (OHIP) or other provincial health plans. Growers prefer that their workers do not make claims for compensation, for fear of having their premiums raised. And Mexicans do not wish to displease their *patrones*, even when they are aware of their right to receive compensation. For the same reason, Mexican workers do not demand public holiday and vacation pay from their employers, even when they realize that they are entitled to receive these benefits by law. Most consider these losses relatively minor compared to what they might lose if they were to confront their employers. They are willing to accept some abuse in exchange for positive letters of evaluation, since without them, their chances of getting their contracts for work in Canada renewed would be very slim.

Earlier chapters have identified two more reasons why Mexicans are always available to work. Since migrant workers leave their families, friends, and communities behind and generally do not form new ones in Canada, they lack the social obligations that might conflict with their commitment to their employers. Finally, their loyalty to their employers is reinforced through the paternalistic relationships between the farmers and their workers. And thus Canadian growers secure labour that cannot be recruited among domestic workers, labour that is relatively cheap, productive, docile, disciplined, and, most importantly, fully committed. Because these unfree workers are available to harvest the crops whenever they are ripe, growers are able to minimize their losses and maintain viable business operations. Without them, many growers would be unable to withstand the losses they would inevitably incur.

The central argument made of this book, therefore, is that Mexican workers have become structurally necessary for Canadian horticulture. This argument departs from other arguments that link the need for foreign labour to the vulnerability of the sectors employing it. Various Canadian researchers (discussed in chapter 4) have argued that domestic workers are not interested in farm jobs, because of low wages and poor working conditions. At the same time it has been suggested that many family farms that are squeezed between the rising costs of farming and the low prices they are forced to charge for their produce could neither raise wages nor improve the working environment.

This line of reasoning is representative of the "segmented labour market" approach, which was discussed in chapter 1 and which associates vulnerable economic sectors with "secondary sector" jobs that cannot be filled by domestic workers and therefore require (im)migrant labour. There are two problems with this argument. The first one, outlined in the introduction, is that it fails to explain when it is necessary to employ offshore workers, as opposed to members of domestic "marginal" groups, such as youth, visible minorities, newly arrived immigrants, and especially women, many of whom may be willing to work for low pay under difficult and dangerous conditions. Second, this argument fails to take social differentiation in the farming industry into consideration. Not all agricultural businesses employing foreign labour are vulnerable family farms. Among the Leamington growers surveyed in this study, many have generated sales and profits that are significantly higher than provincial and national averages. Many growers have expanded their businesses by purchasing more land or greenhouses. Alongside these large greenhouse operations, some family farms are just breaking even or incur losses. Despite the differences between these farms, most depend heavily on offshore labour, since both groups would be harmed if crops were not picked on time. Small farms would suffer if the labour was not available and crops were lost, because they are barely keeping afloat. Yet because of the large scale of their operations, losses of crops caused by a shortage of labour in large greenhouses would be enormous and the consequences for the economic survival of these businesses could be serious. The argument developed in this book is that it is not the vulnerability of farming that makes growers dependent on offshore labour but the need to secure labour that is not only unfree to change jobs but that is available for work on demand. Since such unfree labour does not exist in Canada, it is vital for Canadian agriculture to have access to unfree foreign workers. This conclusion has an important implication for the debate on the adverse effects of guest worker programs.

Critics of guest worker programs have raised concerns about the effects of the importation of labour on domestic workers. First, it has been pointed out that native workers are displaced by contract workers (Craig 1971, 24–33; Griffith 1986). With respect to the British West Indies Temporary Alien Labor Program (BWI) in the United States, for instance, Griffith points out that apple and sugar growers refuse to employ domestic workers despite the fact that "there are U.S. domestic workers who I have experienced work settings similar to those of the sugar and apple camps; 2 have been unemployed during the times of year when apple and sugar growers need labor; 3 have worked for annual incomes less than or equal to those earned by the BWI workers; 4 are located in or near the areas where the apple and sugar growers require them to work; and 5 are experienced in harvests similar or identical to sugar and apple harvests (Griffith 1986, 883)."

Similar arguments were raised by organized labour during the *bracero* years in the United States (Craig 1971). Even though the offshore program in Canada has not received such heated criticism as the two U.S. labour recruitment programs, some concern about the effects of this program on domestic labour have been voiced. A report in 1974 commissioned by the Ontario Federation of Labour (Ward 1974, 57–8) spoke of "thousands of jobless Canadians sleeping in parks, shacks, box cars and railroad sidings," searching for work in Southern Ontario and ready to accept farm employment while "over 3,000 West Indians were employed at assured rates and living conditions." In May of 1991, a city councillor in London, Ontario, suggested that West Indian seasonal workers could be replaced with the city's unemployed or welfare-receiving residents (Cecil and Ebanks 1992, 25). In 1980 Canada Employment and Immigration Commission report expressed a concern that the number of foreign workers authorized to work in Canadian agriculture exceeded the need for supplementary labour (CEIC 1980, 14). Groups representing labour have argued that the dependence of growers on foreign workers delivered by the Seasonal Agricultural Workers Program denies Canadians access to employment. It has been pointed out that the requirement of placing requests for foreign workers eight weeks in advance makes it virtually impossible for the CEIC to recruit the domestic workers who are available for short-term work just two to four weeks in advance of the harvest periods (3).

As I have argued, cheap workers who are unfree to change jobs but who are fully committed to their employers are not available in Canada. Canadian workers who accept farm jobs are free to quit and to take time off in the middle of the harvest. Offshore workers available to work on demand supplement Canadian labour employed in the industry. In fact, by providing a safety net to the growers, offshore workers have sustained many farming businesses, and many jobs have been saved and new jobs have been created as a result.

According to FARMS (1995), the Ontario horticulture industry required a labour force of 99,876 workers in 1995, but only 90 percent of these jobs were taken by Canadians, leaving 9,876 vacancies to be filled by offshore workers. At the same time, Ontario farmers invested $626 million in seedstocks, chemicals, equipment, and other goods and services and thus supported approximately 2,500 jobs on the supply side of the industry. They also contributed to the creation of 49,938 jobs in the food processing industry, which was staffed predominantly by Canadian labour. Thus, each farm worker in horticulture supported 2.6 jobs in the supply and processing sectors, and if the 9,876 jobs in the Ontario horticultural industry had not been filled by offshore workers, 25,678 jobs in other sectors would have been lost.

In the last few years, the greenhouse industry has become a year-round business in Leamington, but Mexican workers are allowed to work in Canada only for a maximum of eight months. When they leave by (November 1), only local workers are employed to clean the houses and to plant new crops. Some of them continue working on the farms after the Mexicans return. Furthermore, the boom in the greenhouse industry, which is related to the employment of cheap and reliable offshore workers, has created many jobs in construction that have been filled exclusively by local workers.

This situation is not unique. Various researchers have argued that domestic workers benefit indirectly from the employment of foreign workers, insofar as it prevents labour bottlenecks and sustains growth (e.g., Miller 1986, 747). As Castles and Kosack observe with respect to the economic sectors that rely on foreign the labour due to labour shortages they experience, "If these branches were crippled by labour shortages, the resulting bottlenecks would affect many other industries" (1985, 399–400).

Critics of guest worker programs have also pointed out that seasonal migration puts downward pressure on the average wage (Craig 1971, 24–33; Verhaeren 1986).* I would argue, however, that a guest worker program is likely to adversely affect wages under two conditions: when employment standards are either

* Critics of guest worker programs also argue that many seasonal workers do not return home after their contract expires. Whether guest worker programs contribute to the rise in undocumented population in a host country is discussed in Basok (2000b).

unspecified or poorly enforced by program administrators, and when there is a pool of domestic workers interested in the jobs offered to foreign seasonal workers. The first condition does not exist in Canada. As discussed in chapter 2, the Seasonal Agricultural Workers' Program outlines a set of conditions under which offshore workers are to be employed. In Ontario' offshore workers are covered by provincial Employment Standards Act, and growers, by and large, are bound by the act. In fact, it has been pointed out that the employment standards established for the employment of foreign workers have positively influenced working conditions for Canadian farm workers by standardizing wages and other employment-related conditions (CEIC 1980, 8). Ward (1974) reports that in the 1970s employment conditions for Caribbean offshore workers were better than for Canadian farm workers, that many domestic workers were poorly housed and underpaid, and that almost half the Ontario farm workers received less than a minimum wage. Among other things, the practice of "banking" workers' earnings until the end of the season created much hardship for workers and their families who had no savings at the beginning of the harvest season. Workers who left their jobs before the end of the season never received the wages "banked" for them (22–3). With time, however, the working conditions for local workers have been equalized with those extended to the guest workers.

With respect to the second condition, the argument gets more complicated. On the one hand, as chapter 4 illustrated, not many Canadian workers are interested in taking farm jobs for the wages that farmers have been willing to pay – generally between seven and nine dollars per hour. Several growers have expressed doubts that they would find more committed and reliable workers even if they paid them a few dollars more. Yet if the wages went up significantly – to fifteen or twenty dollars per hour, growers would probably be able to retain many more workers. While many of the growers interviewed argued that they could not afford to pay such high wages, some among them probably could. However, since even at a significantly higher level of remuneration growers cannot expect full commitment from local workers, who are generally reluctant to become "chained" to their jobs, from the growers' point of view, wage increases to local workers would be considered a bad investment. The availability of offshore workers may prevent those growers who could afford it from increasing their workers' pay, and in this sense the offshore program does push local wages down. But there are also other reasons why wages are kept at existing levels. The good fortune that Leamington greenhouse growers have experienced in the last decade may not be long-lasting. They are beginning to face some competition from Dutch, U.S., and Mexican greenhouse growers. Several growers claimed that sales for 1999 were considerably lower than those reported for 1998. At the turn of the century greenhouse growers were hit by two disasters threatening to shrink their profits – mounting fuel prices and a trade dispute with U.S. tomato growers, who accused their Canadian competitors of selling tomatoes cheaper in the United States than in Canada and below the cost of production (Welch 2001b; Hill 2001). The insecurity of sales created by weather and market fluctuations affecting every grower, even the most successful among them, would make it highly unlikely that they would be willing to substantially increase their costs of operation by raising wages. And if they did, they would risk ruining their businesses. Furthermore, even if some larger businesses decided to raise wages in order to retain local workers, the effect of this initiative on smaller farms could be so devastating that it would force many of them out of business, leading to increases in unemployment not only among the farm operators and members of their households but also among the local workers they employ.

Postscript

In the middle of May 2001 I received a phone call from United Farm Workers, who were interested in my research on Mexican seasonal workers. This interest was triggered by a two-day walk-out organized by the workers employed at one big greenhouse in Leamington. I had heard about the walk-out from Mary Agnes Welch, a *Windsor Star* correspondent, who had asked me to translate a petition written by the rebellious workers a week before the phone call from United Farm Workers. I was told that on Sunday a group of people from Toronto, including representatives from the United Farm Workers, the Canadian Labour Congress, student activists, journalists, and interpreters were to meet with Mexican workers in Leamington after the service at St Michael's church. A local MP was also invited to attend, and Mary Agnes Welch was going there on her quest to understand the causes and consequences of the walk-out. I decided to join them.

What I experienced that Sunday afternoon was a replay of the events I discussed in the preface to this book. As before, Mexican workers were anxious to tell anyone who was willing to listen about the abuse they had experienced on Canadian farms. They talked about accidents at work and their uncertainty about compensation insurance coverage. They described the deplorable living conditions some of them experienced. They talked about long hours of work without overtime pay. Some were frustrated about not having received their income tax return from the Mexican consulate, and they were very critical of representatives of the Mexican consulate in general. They were upset that unemployment insurance premiums were deducted from their pay without them benefitting from this insurance policy. And they were uncertain about whether they could ever collect their Canadian pension. I had a feeling of *déja vu.*

What also seemed painfully familiar was the reluctance of some workers to openly criticize their *patrones* for fear of possible repercussions. Mary Agnes Welch and I joined forces in trying to locate and interview workers from the troubled greenhouse. It was not easy. One person who was pointed out to us refused to tell us anything. Two others did talk to us, but without much enthusiasm. We found out that some workers on that farm were pushed around and constantly denigrated by their foreman and that, consequently, they walked out demanding that they be treated with respect. The management asked other Mexican workers to work long hours in order to complete the tasks left unattended by the workers who had walked out. Under different circumstances these workers would have welcomed the opportunity to work extra hours, but in solidarity with the others they also walked out.* According to the two workers with whom we spoke, by then the problem that had led to the walk-out had been resolved, even though about twenty "ringleaders" had been sent back to Mexico and the foreman who had driven the workers to take action was still employed by the company. The two workers also prefaced their conversation with us by saying that they had walked out in solidarity with other workers and not because they had experienced problems on that farm.

And yet, unlike my aborted activism in 1997, the events unfolding on that day in May 2001 gave me hope. The United Farm Workers had become concerned with the plight of foreign farm workers in Canada and had set the objective of documenting and exposing the abuse and injustice these workers experience and demanding improvements in their treatment from those responsible for the program. They wanted to see living quarters regularly inspected, unemployment premium deductions lifted, exposure to dangerous pesticides reduced, and abusive employers sanctioned.

* See Mary Agnes Welch's account of the events that led to the strike in Welch (2001a, b).

Was this activism likely to lead to the demise of the Seasonal Agricultural Workers Program? As I have argued in this book, it is not the low cost of Mexican workers that makes them vital to Canadian growers but the fact that they are available for work on demand. Improvements in their living conditions, paid public holidays, and a healthier working environment will create additional costs for growers. And while for some growers the additional financial burden may make it difficult to stay in business, most can afford these extra costs. Their losses would be much greater if they did not have access to this "captive" labour force. With their working and living conditions improved, Mexican workers are likely to feel even more loyal to their *patrones* than they do already, and from that point of view these improvements would be an investment well spent.

IN THIS POSTSCRIPT, THE AUTHOR CLEARLY TAKES THE STAND OF GROWERS, ADVISING THEM TO IMPROVE THEIR WKS CONDITIONS SO AS TO KEEP HAVING ACCESS TO THIS POOL OF UNFREE LABOUR!

Glossary

banco-a-banco	bank transfer
bracero	a Mexican guest worker employed in the United States between 1942 and 1964
castigado/a	punished
coyote	a middleman who transports undocumented migrants over the Mexico-U.S. border
de mojado	illegally (literally, wetback)
Dios primero	if God is willing (a frequently used expression)
ejido	rural community
gritos	proclamations
jornalero	day worker, peon
licenciado	an official employed in government offices (literally, someone with a bachelor's degree)
mariachis	musicians playing and singing traditional serenades
migra	immigration authorities
mordida	bribe
oficinas	offices (may refer to the Mexican Ministry of Labour and Social Planning or to other offices)
paisanos	individuals from the same region
patrón/patrones	employer/employers
primaria	primary school
rancho	village
preparatoria	high school

Secretaría	an abbreviation for the Ministry of Labour and Social Planning
secundaria	secondary school
tabique	cinder blocks
telesecundaria	distance education secondary school

Bibliography

Alejandre, Jesus Arroyo, Adrian de Leon Arias, and Basilia Valenzuela Varela. 1991. "Patterns of Migration and Regional Development in the State of Jalisco, Mexico." In *Regional and Sectoral Development in Mexico as Alternatives to Migration*, edited by Sergio Diaz-Briquets and Sidney Weintraub. Boulder, CO: Westview Press.

Alston, Lee, and Joseph Ferrie. 1993. "Paternalism in Agricultural Labor. Contracts in the U.S. South: Implications for the Growth of the Welfare State." *American Economic Review* 83: 852–76.

Anderson, R.W., and R.D. Daniel. 1977. "Canada"s Fruit and Vegetable Processing Industry." *Canadian Farm Economics* 12 (4): 10–16.

Argüello, Francisco. 1993. "Experiencias Migratorias de Campesinos de Guanajuato en Canadá." *Regiones: Revista Interdisciplinaria en Estudios Regionales* 1 (1): 89–97.

Barrón, Antonieta. 1999. "Mexican Women on the Move: Migrant Workers in Mexico and Canada." In *Women Working the NAFTA Food Chain: Women, Food and Globalization*, edited by Deborah Barndt. Toronto: Second Story Press.

Basok, Tanya. 1999. "Free to Be Unfree: Mexican Guest Workers in Canada." *Labour, Capital and Society* 32 (2): 192–221.

– 2000a. "Obstacles to Productive Investment: Mexican Farm Workers in Canada," *International Migration Review* 34 (1): 79–97.

– 2000b. "He Came, He Saw, He ... Stayed. Guest Worker Programs and the Issue of Non-Return," *International Migration* 38 (2): 215–38.

Berger, Suzanne, and Michael J. Piore. 1980. *Dualism and Discontinuity in Industrial Societies*. Cambridge: Cambridge University Press.

Bezaire, Ernest. 1965. *Agricultural Labour in Southwestern Ontario*. Leamington, ON: Essex County Associated Growers.

Bluestone, Barry, and Bennett Harrison. 1982. *The Deindustrialization of America: Plant Closings, Community Abandonment, and the Dismantling of Basic Industries*. New York: Basic Books.

Bolaria, B. Singh. 1988. "The Health Effects of Powerlessness: The Case of Immigrant Farm Labour." In *The Political Economy of Agriculture in Western Canada*, edited by G.S. Basran and D.A. Hay. Toronto: Garamond Press.

– 1992. "Farm Labour, Work Conditions, and Health Risks." In *Rural Sociology in Canada*, edited by David Hay and Gurcharn Basran. Toronto: Oxford University Press.

Bolaria, B. S., Harley Dickinson, and Terry Wotherspoon. 1991. "Rural Issues and Problems." In *Social Issues and Contradictions in Canadian Society*, edited by B. Singh Bolaria. Toronto: Harcourt Brace Jovanovich.

Bonacich, Edna. 1972. "A Theory of Ethnic Antagonism: The Split Labour Market." *American Sociological Review* 37: 547–59.

Boston, Thomas. 1990. "Segmented Labor Markets: New Evidence from a Study of Four Race-Gender Groups." *Industrial and Labor Relations Review* 44 (1): 99–115.

Boswell, Terry, and John Brueggemann. 2000. "Labor Market Segmentation and the Cultural Division of Labor in the Copper Mining Industry, 1880–1920." *Research in Social Movements, Conflicts, and Change* 22: 193–217.

Calavita, Kitty. 1992. *Inside the State: The Bracero Program, Immigration, and the I.N.S.* New York: Routledge.

Canada. Department of Manpower and Immigration. 1973. *The Seasonal Farm Labour Situation in Southwestern Ontario*.

Carroll, William, and Rennie Warburton. 1991. "Capital, Labour, and the State: The Future of the Labour Movement." In *Social Issues and Contradictions in Canadian Society*, edited by B. Singh Bolaria. Toronto: Harcourt Brace Jovanovich.

Castells, M. 1979. "Immigrant Workers and Class Struggles in Advanced Capitalism: The Western Europen Experience." In *Peasants and Proletarians: The Struggles of Third World Workers*, edited by Robin Cohen et. al. New York: Monthly Review Press.

Castles, S., and G. Kosack. 1985. *Immigrant Workers and Class Structure in Western Europe*. Oxford: Oxford University Press, 1973. Reprint.

Castles, S., with Heather Booth and Tina Wallace. 1984. *Here for Good: Western Europe's New Ethnic Minorities*. London: Pluto Press.

Cecil R.G., and G.E. Ebanks. 1991. "The Human Condition of West Indian Migrant Farm Labour in Southwestern Ontario." *International Migration* 29 (3): 389–405.

– 1992. "The Caribbean Migrant Farm Worker Programme in Ontario: Seasonal Expansion of West Indian Economic Spaces." *International Migration* 30 (1): 19–37.

CEIC (Canada Employment and Immigration Commission). 1980. *Review of Agricultural Programs, 1980*. Ottawa.

Cohen, Robin. 1987. *The New Helots: Migrants in the International Division of Labour.* Aldershot, England: Avebury.

Colby, Catherine. 1997. *From Oaxaca to Ontario: Mexican Contract Labor in Canada and the Impact at Home*. Davis, CA: The California Institute for Rural Studies.

Cornelius, W.A. 1991. "Labour Migration to the United States: Development Outcomes and Alternatives in Mexican Sending Communities." In *Regional and Sectoral Development in Mexico as Alternatives to Migration*, edited by Sergio Diaz-Briquets and Sidney Weintraub. Boulder, CO: Westview Press.

– 1998. "The Structural Embeddedness of Demand for Mexican Immigrant Labour: New Evidence from California." In *Crossings: Mexican Immigration in Interdisciplinary Perspectives*, edited by Marcelo Suarez-Orozco. Cambridge, MA: David Rockefeller Centre for Latin American Studies, Harvard University Press.

Cornies, Larry. 1977a. "Portuguese Community." *Leamington Post*, 2 February.

– 1977b. "German-Swabian Community." *Leamington Post*, 9 February.

– 1977c. "Italian Community." *Leamington Post*, 2 March.

Craig, Richard. 1971. *The Bracero Program. Interest Groups and Foreign Policy.* Austin, TX: University of Texas Press.

Dawson, Donald A., and David Freshwater. 1975. *Hired Farm Labour in Canada*. Food Prices Review Board. Mimeographed.

Delgado, Gary. 1983. "Organizing Undocumented Workers." *Social Policy* 13: 26–9.

Driedger, N.N. 1972. *The Leamington United Mennonite Church: Establishment and Development, 1925–1972*. Altona, MB: D. N. Friesen.

Durand, Jorge, and Douglas S. Massey. 1992. "Mexican Migration to the United States: A Critical Review." *Latin American Research Review* 27: 3–42.

Dussel Peters, Enrique. 1998. "Recent Structural Changes in Mexico"s Economy: A Preliminary Analysis of Some Sources of Mexican Migration to the United States." In *Crossings: Mexican Immigration in Interdisciplinary Perspectives*, edited by Marcelo M. Suárez-Orozco. Cambridge, MA: David Rockefeller Centre for Latin American Studies, Harvard University Press.

Edwards, Richard C. 1973. "The Social Relations of Production in the Firm and Labour Market Structure." In *Labour Market Segmentation*, edited by Richard Edwards, Michael Reich, and David M. Gordon. Lexington, MA: D.C. Heath and Company.

Edwards, Richard C., Michael Reich, and David M. Gordon, eds. 1973. *Labour Market Segmentation*, Lexington, MA: D.C. Heath and Company.

FARMS. 1995. *The Quest for a Reliable Workforce in the Horticulture Industry: Reliable Workers, Regardless of Source*. Mississauga, ON. Mimeographed.

– 1996. *1996 Field Study, A Follow-Up to the Quest for a Reliable Workforce in the Horticulture Industry: Reliable Workers, Regardless of Source*. Mississauga, Ontario. Mimeographed.

– 1999. *Employer Information Package*. Mississauga, Ontario. Mimeographed.

Fernández Kelly, María Patricia. 1985. "Contemporary Production and the New International Division of Labor." In *The Americas in the New International Division of Labor*, edited by Steven E. Sanderson. New York: Holmes and Meier.

Galarza, Ernesto. 1964. *Merchants of Labor: The Mexican Bracero Story*. Sage Yearbook in Politics and Public Policy. Santa Barbara, CA: McNally and Loftin.

Ghorayshi, Parvin. 1986. "The Identification of Capitalist Farms: Theoretical and Methodological Considerations." *Sociologia Ruralis* 26 (2): 146–59.

González Baker, Susan, Frank D. Bean, Augustin Escobar Latapi, and Sidney Weintraub. 1998. "U.S. Immigration Policies and Trends: The Growing Importance of Migration from Mexico." In *Crossings: Mexican Immigration in Interdisciplinary Perspective*, edited by Marcelo M. Suárez-Orozco. Cambridge, MA: David Rockefeller Cente for Latin American Studies, Harvard University Press.

Gordon, David, Richard Edwards, and Michael Reich. 1982. *Segmented Work, Divided Workers: The Historical Transformation of Labour in the United States*. Cambridge: Cambridge University Press.

Green, Duncan. 1995. *Silent Revolution: The Rise of Market Economics in Latin America*. London: Cassell.

Green, Susan. 1983. "Silicon Valley's Women Workers: A Theoretical Analysis Market." In *Women, Men and the International Division of Labor*, edited by June Nash and María Patricia Fernández-Kelly. Albany, NY: State University of New York Press.

"Greenhouse Growers Gain by Diagnostic Program." *Agvance* 6 (4).

Greenhouse Sector. 1996. *Leamington and Area Greenhouse Growers' Directory*. http:/res.agr.ca/harrow/grv/ind.htm.

Griffith, David. 1986. "Peasants in Reserve: Temporary West Indian Labor in the U.S. Farm Labor Market." *International Migration Review* 20 (4): 875–98.

Harris, Nigel. 1995. *The New Untouchables: Immigration and the New World Worker*. London: I.B. Tauris Publishers.

Harrison, Bennett. 1994. *Lean and Mean: The Changing Landscape of Corporate Power in the Age of Flexibility*. New York: Basic Books.

Harrison, Bennett, and Andrew Sum. 1979. "The Theory of 'Dual,' or Segmented, Labor Markets." *Journal of Economic Issues* 8 (3): 687–703.

Haythorne, George. 1960. *Labor in Canadian Agriculture*. Cambridge, MA: Harvard University Press.

Hill, Sharon. 1999a. "Greenhouse Growth Spawns Industry Study." *Windsor Star*, 14 September.

– 1999b. "Greenhouse Industry Puts Out Welcome Mat," *Windsor Star*, 15 September.

– 2000a. "Greenhouse Control Needed." *Windsor Star*, 28 March.

– 2000b. "Town Imposes Ban on New Greenhouses." *Windsor Star*, 13 June.

– 2001. "Dicey Future for Tomato Growers." *Windsor Star*, 11 May.

Howell, Frances. 1982. "A Split Labor Market: Mexican Farm Workers in the Southwest." *Sociological Inquiry* 52 (2): 132–40.

Jaret, Charles. 1991. "Recent Structural Change and U.S. Urban Ethnic Minorities." *Journal of Urban Affairs* 13 (3): 307–36.

Jansen, William. 1988. "The Mennonites from Mexico in Ontario: Who Are They?" Unpublished paper presented in Aylmer, Ontario, 6 October.

Jones, R.C., ed. 1984. *Patterns of Undocumented Migration: Mexico and the U.S.* Totowa, NJ: Rowman and Allenheld.

Katz, Naomi, and David Kemnitzer. 1983. "Fast Forward: The Internationalization of Silicon Valley." In *Women, Men, and the International Division of Labour*, edited by June Nash and Maria Patricia Fernandez Kelly. Albany, NY: State University of New York Press.

Khosla, Shalin. 1998. *The Ontario Greenhouse Vegetable Industry*. Harrow, ON: OMAFRA, Greenhouse and Vegetable Crops Research Centre.

Kliewer, Victor. 1997. *The Mennonites in Essex and Kent Counties, Ontario: An Introduction*. Leamington, ON: Essex-Kent Mennonite Historical Association.

Knowles, Kimberley. 1997. "The Seasonal Agricultural Workers Program in Ontario: From the Perspective of Jamaican Migrants." MA thesis, University of Guelph.

Lackrey, Cynthia. 1998. "Harrow Greenhouse Research Leads the World." *Voice of the Farmer*, 7 July.

Laliberte, Ron, and Vic Satzewich. 1999. "Native Migrant Labour in the Southern Alberta Sugarbeet Industry: Coercion and Paternalism in the Recruitment of Labour." *Canadian Review of Sociology and Anthropology* 36 (1): 65–85.

Larkin, Sherrie. 1990. *West Indian Workers and Ontario Farmers: The Reciprocal Construction of a Divided World*. MA thesis, University of Western Ontario.

Lipsig-Mammé. 1987. "Organizing Women in the Clothing Trades: Homework and the 1983 Garment Strike in Canada." *Studies in Political Economy* 22: 41–71.

Marx, Karl.1976. *Capital*. Vol 1. Harmondsworth, England: Penguin.

Massey, Douglas S., and Lawrence C. Basem. 1992. "Determinants of Savings, Remittances, and Spending Patterns among U.S. Migrants in Four Mexican Communities." *Sociological Inquiry* 62 (2): 185–207.

Massey, D.S., and K.E. Espinosa. 1997. "What"s Driving Mexico-U.S. Migration? A Theoretical, Empirical, and Policy Analysis." *American Journal of Sociology* 102 (4): 939–99.

Miles, Robert. 1987. *Capitalism and Unfree Labour. Anomaly or Necessity?* London: Tavistock Publications.

Miller, Mark. 1986. Introduction to *International Migration Review, Special Issue. Temporary Worker Programs: Mechanisms, Conditions, Consequences* 20 (4): 740–57.

Mitchell, Don. 1975. *The Politics of Food*. Toronto: James Lorimer.

Morokvasic, Mirjana. 1984. "Birds of Passage Are Also Women." *International Migration Review* 18 (4): 886–907.

Morrison, Neil. 1954. *Garden Gateway to Canada: One Hundred Years of Windsor and Essex County, 1854–1954*. Toronto: The Ryerson Press.

Nash, June. 1983. "The Impact of the Changing International Division of Labor on Different Sectors of the Labor Force." In *Women, Men, and the International Division of Labour*, edited by June Nash and Maria Patricia Fernandez Kelly. Albany, NY: State University of New York Press.

Nash, June, and María Patricia Fernández-Kelly, eds. 1983. *Women, Men and the International Division of Labor.* Albany, NY: State University of New York Press.

OGVPMB (Ontario Greenhouse Vegetable Producers' Marketing Board). 1998. *1998 Annual Report.* Mimeographed.

Panitch, Leo, and Donald Swartz. 1988. *The Assault on Trade Union Freedoms: From Consent to Coercion Revisited.* Toronto: Garamond Press.

Parr, J. 1985. "Hired Men: Ontario Wage Labour in Historical Perspectives." *Labour/Le Travail* 15: 91–103.

Pentland, H.C. 1981. *Labour and Capital in Canada, 1650–1860.* Toronto: James Lorimer.

Phizacklea, A., ed. 1983. *One-Way Ticket? Migration and Female Labour.* London: Routledge and Kegan Paul.

Phizaclea A., and R. Miles. 1980. *Labour and Racism.* London: Routledge and Kegan Paul.

Piore, Michael. 1973. "Notes for a Theory of Labor Market Stratification." In *Labor Market Segmentation.* edited by Richard C. Edwards, Michael Reich, and David Gordon. Lexington, MA: D.C. Heath.

– 1979. *Birds of Passage: Migrant Labor and Industrial Societies.* Cambridge: Cambridge University Press.

Portes, A. 1978. "Toward a Structural Analysis of Illegal Undocumented. Immigration." *International Migration Review* 12: 469–84.

Preibisch, Kerry. 1998. "La tierra de los no libres: migración temporal México-Canadá y dos campos de reestructuración económica neoliberal." Paper presented at the Latin American Studies Association Twenty-First International Congress, Chicago, 24–26 September.

Ramirez, Miguel D. 1989. *Mexico's Economic Crisis: Its Origins and Consequences.* New York: Praeger.

Reichert, Joshua S. 1981. "The Migrant Syndrome: Seasonal U.S. Wage Labour and Rural Development in Central Mexico." *Human Organization* 40: 56–66.

Reichert, Joshua, and Douglas Massey. 1982. "Guestworker Programs: Evidence from Europe and the United States and Some Implications for U.S. Policy." *Population Research and Policy Review* 1: 1–17.

Rempel H., and R.A. Lobdell. 1978. "The Role of Urban-to-Rural Remittances in Rural Development." *Journal of Development Studies* 14 (3): 324–41.

Russell, Philip. 1994. *Mexico under Salinas.* Austin, TX: Mexico Resource Center.

Sanderson, George. 1974. "The Sweatshop Legacy: Still with Us in 1974." *Labour Gazette* 74: 400–17.

Santos, Nelson. 1998. "Investment in Research Opens New Doors. Area Greenhouse Growers to Benefit." *Kingsville Reporter*, 7 July.

Sassen, S. 1988. *The Mobility of Labour and Capital*. Cambridge: Cambridge University Press.

– 1996. "New Employment Regimes in Cities: The Impact on Immigrant Workers." *New Community* 22 (4): 579–94.

Sassen-Koob, Saskia. 1985. "Capital Mobility and Labor Migration." In *The Americas in the New International Division of Labor*, edited by Steven E. Sanderson. New York: Holmes and Meier.

Satzewich, Vic. 1991. *Racism and the Incorporation of Foreign Labour: Farm Labour Migration to Canada since 1945*. London: Routledge.

Sawatzky, Harry L. 1971. *They Sought a Country: Mennonite Colonization in Mexico*. Berkley, CA: University of California Press.

Schmidt, Doug. 1999. "Growers Dubious about Workfare Farm Labour." *Windsor Star*, 2 September.

Scott, James. 1985. *Weapons of the Weak: Everyday Forms of Peasant Resistance*. New Haven, CT: Yale University Press.

Sheik, Joe. 1987. *Leamington: Pioneer Village to Agricultural Center*. Leamington, ON.

Shields, John. 1988. "The Capitalist State and Class Struggle in the Fruit and Vegetable Industry in British Columbia." In *The Political Economy of Agriculture in Western Canada*, edited by G.S. Basran and D.A. Hay. Toronto: Garamond Press.

– 1992. "The Capitalist State and Farm Labour Policy." In *Rural Sociology in Canada*, edited by David Hay and Gurcharn Basran. Toronto: Oxford University Press.

Smart, Josephine. 1997. "Borrowed Men on Borrowed Time: Globalization, Labour Migration, and Local Economies in Alberta." *Canadian Journal of Regional Science* 20 (1–2): 141–56.

Smit, Barry, Tom Johnston, and Robert Morse. 1984. *Employment and Labour Turnover in Agriculture: A Case Study of Flue-Cured Tobacco Farms in Southern Ontario*. Occasional Papers in Geography No. 5. Department of Geography, University of Guelph.

Snell, Frances. 1974. *Leamington's Heritage, 1874–1974*. Leamington, ON: The Town of Leamington.

Stalker, Peter. 2000. *Workers without Frontiers: The Impact of Globalization on International Migration*. Boulder, CO: Lynne Rienner Publishers.

Stevens, R. Wm. 1996. *Industrial Adjustment Service Committee: Final Report, Executive Summary.* Guelph, ON: The Ontario Greenhouse Vegetable Growers' Marketing Board.

Stultz, Erma. 1987. "Organizing the Unorganized Farmworkers in Ontario." In *Working People and Hard Times: Canadian Perspectives*, edited by Robert Argue, Charlene Gannagé, and D.W. Livingstone. Toronto: Garamond Press.

Sullivan, Glenn, and Kevin Garleb. 1996. *Summary Report: Ontario Greenhouse Vegetable Strategic Planning Initiative.* Mimeographed.

Tatroff, Daniel. 1994. "Fields of Fear: Picking for a Living in B.C." *Our Times* 13 (6): 22–7.

Terry, James. 1988. *The Political Economy of Migrant Farm Labor and the Farmworker Movement in the Midwest.* PHD dissertation, Purdue University.

Tuddenham, Edward. 1985. "The False Promise of Legalized Immigration in Agriculture." In *In Defense of the Alien.* Vol. 8, *Immigration Enforcement, Employment Policy, Migrant Rights and Refugee Movements*, edited by Lydio Tomasi. New York: Centre for Migration Studies.

Verduzco, Gustavo. 2000. "El Programa de Trabajadores Agrícolas Mexicanos con Canadá: aprendizaje de una nueva experiencia." In *Canada: un estado posmoderno*, edited by Maria T. Gutierrez Hoces. Mexico City, Mexico: Asociación Mexicana de Estudios Canadienses.

Verhaeren, Raphael-Emanuel. 1986. "The Role of Foreign Workers in the Seasonal Fluctuations of the French Economy." *International Migration Review* 20 (4): 856–74.

Waddoups, Jeffrey, and Djeto Assane. 1993. "Mobility and Gender in a Segmented Market: A Closer Look." *American Journal of Economics and Sociology* 52 (4): 399–412.

Waldinger, Roger. 1985 "The Garment Industry in New York City." In *Hispanics in the U.S. Economy*, edited by George Borjas and Marta Tienda. New York: Academic Press.

– 1997. *Social Capital or Social Closure? Immigrant Networks in the Labor Market.* Los Angeles, CA: Lewis Center for Regional Policy Studies, University of California.

Wall, Ellen. 1992. "Personal Labour Relations and Ethnicity in Ontario Agriculture." In *Deconstructing a Nation: Immigration, Multiculturalism and Racism in '90s Canada*, edited by Vic Satzewich. Halifax, NS: Fernwood Publishing.

– 1994. "Farm Labour Markets and the Structure of Agriculture." *Canadian Review of Sociology and Anthropology* 31 (1): 64–81.

Ward, B. 1974. *Harvest of Concern: Conditions in Farming and Problems of Farm Labour in Ontario*. Ottawa: Ontario Federation of Labour.

Warnock, John. 1978. *Profit Hungry: The Food Industry in Canada*. Vancouver: New Star Books.

– 1995. *Other Mexico: The North American Triangle Completed*. Montreal: Black Rose Books.

Warrian, Peter. 1987. "Trade Unions and the New International Division of Labour." In *Working People and Hard Times: Canadian Perspectives*, edited by Robert Argue, Charlene Gannagé, and D.W. Livingstone. Toronto: Garamond Press.

Welch, Mary A. 2000a. "Greenhouses Dead without Cheap Labour." *Windsor Star*, 2 November.

– 2000b. "The Migrant Code: 'Don't Ask Questions.'" *Windsor Star*, 3 November.

– 2001a. "Migrants Air Workplace Complaints." *Windsor Star*, 22 May.

– 2001b. "Migrant Workers' Claims 'Unfair.'" *Windsor Star*, 26 May.

Whitfield, G., and A.P. Papadopoulos. 1999. *Introduction to the Greenhouse Vegetable Industry*, http:/res.agr.ca/harrow/hrcghar.htm.

Wilkinson, Doris. 1991. "The Segmented Labor Market and African American Women from 1890–1960: A Social History Interpretation." *Research in Race and Ethnic Relations* 6: 85–104.

Winson, Anthony. 1990. "Capitalist Coordination of Agriculture: Food Processing Firms and Farming in Central Canada." *Rural Sociology* 55 (3): 376–94.

– 1992. *The Intimate Commodity: Food and the Development of the Agro-Industrial Complex in Canada*. Garamond Press.

– 1996. "In Search of the Part-Time Capitalist Farmer: Labour Use and Farm Structure in Central Canada." *Canadian Review of Sociology and Anthropology* 33 (1): 89–110.

Wood, C.H., and T.L. McCoy. 1985. "Caribbean Cane Cutters in Florida: Implications for the Study of the Internalization of Labor." In *The Americas in the New International Division of Labor*, edited by Steven Sanderson. New York: Holmes and Meier.

Index